Essentials of Geophysical Data Proces

This textbook provides a concise introduction to geophysical data processing – including many of the techniques associated with the general field of time series analysis – for advanced students, researchers, and professionals. The treatment begins with calculus before transitioning to discrete time series via the sampling theorem, discussions of aliasing, the use of complex sinusoids, the development of the Discrete Fourier Transform from the Fourier series, and an overview of linear digital filter types and descriptions. Aimed at senior undergraduate and graduate students in geophysics, environmental science, and engineering with no previous background in linear algebra, probability, statistics or Fourier transforms, this textbook draws scenarios and data sets from across the world of geophysics and shows how data processing techniques can be applied to real-world problems using detailed examples, illustrations, and exercises. Exercises are mostly computational in nature and may be completed using MATLAB or a computing environment with similar capabilities.

Clark R. Wilson is the Carlton Centennial Professor of Geophysics at the University of Texas at Austin. After undergraduate studies in physics at the University of California, San Diego, and graduate work in geophysics, involving a Masters degree and PhD, at Scripps Institution of Oceanography, he joined the faculty of the Department of Geological Sciences at the University of Texas at Austin. His research has ranged over diverse fields including applied seismology, space geodesy, and hydrology. He has twice served as Department Chair, and also spent three years at the NASA headquarters overseeing programs in geodynamics and potential fields.

Essentials of Geophysical Data Processing

CLARK R. WILSON

University of Texas at Austin

CAMBRIDGE
UNIVERSITY PRESS

CAMBRIDGE
UNIVERSITY PRESS

University Printing House, Cambridge CB2 8BS, United Kingdom

One Liberty Plaza, 20th Floor, New York, NY 10006, USA

477 Williamstown Road, Port Melbourne, VIC 3207, Australia

314–321, 3rd Floor, Plot 3, Splendor Forum, Jasola District Centre, New Delhi – 110025, India

103 Penang Road, #05–06/07, Visioncrest Commercial, Singapore 238467

Cambridge University Press is part of the University of Cambridge.

It furthers the University's mission by disseminating knowledge in the pursuit of education, learning, and research at the highest international levels of excellence.

www.cambridge.org
Information on this title: www.cambridge.org/9781108931007
DOI: 10.1017/9781108939690

First published 2022

Printed in the United Kingdom by TJ Books Limited, Padstow Cornwall

A catalogue record for this publication is available from the British Library.

Library of Congress Cataloging-in-Publication Data
Names: Wilson, Clark R., 1949- author.
Title: Essentials of geophysical data processing / Clark R. Wilson,
University of Texas at Austin.
Description: Cambridge, United Kingdom ; New York, NY : Cambridge
University Press, [2022] | Includes bibliographical references and index.
Identifiers: LCCN 2021038890 | ISBN 9781108931007 (paperback)
Subjects: LCSH: Geophysics–Data processing. | Geophysics. | BISAC: SCIENCE /
Physics / Geophysics
Classification: LCC QC801 .W55 2022 | DDC 550.285–dc23
LC record available at https://lccn.loc.gov/2021038890

ISBN 978-1-108-93100-7 Paperback

Contents

Preface   *page* ix

1 An Introduction with Geophysical Time Series Examples   1
  1.1   Global Mean Sea Level   1
  1.2   Stream Discharge   2
  1.3   Eastern Pacific Sea Level   4
  1.4   El Nino Southern Oscillation (ENSO) Index   4
  1.5   Lake Vostok Ice Core Temperature History   6
  1.6   Hector Mines Earthquake Seismograms   7
  1.7   Simulated Seismograms, White Noise, and Computing Environments   9
  1.8   Chapter Summary   9

2 Analog Signals and Digital Time Series   11
  2.1   Digital Time Series Notation   11
  2.2   Digitizing Analog Signals   12
  2.3   Undersampling and Aliasing   14
  2.4   Time Series Statistics   15
      2.4.1   Mean or Average Value   15
      2.4.2   Variance and Standard Deviation   16
      2.4.3   Autocorrelation   17
  2.5   Numerical Representation of Samples   19
  2.6   Decibels   20
  2.7   Applications to Digital Audio Recording   21
  2.8   Chapter Summary   22
  Exercises   23

3 Sinusoids and Fourier Series   25
  3.1   Sinusoids   25
  3.2   Fourier Series   26
  3.3   Partial Fourier Sums   29
  3.4   Complex Numbers   29
  3.5   Complex Sinusoids   31
  3.6   Chapter Summary   33
  Exercises   34

4 The Discrete Fourier Transform   37
  4.1   Fourier Series in Complex Notation   37
  4.2   From Fourier Series to DFT   38

4.3 Frequency and Time Ordering 39
4.4 DFT Normalization Conventions 40
4.5 Sinusoidal Coefficients of a Climate Time Series 40
4.6 FFT Algorithms 42
4.7 Zero-Padding and Interpolation 43
4.8 DFT Interpolation Example 44
4.9 Analytic Signal Computation and Application to Measuring Surface Wave Dispersion 44
4.10 Chapter Summary 47
Exercises 48

5 Linear Systems and Digital Filters 50
5.1 Linear Filter Equations 50
5.2 Discrete Convolution 51
5.3 Correlation 53
5.4 Convolution Matrices 53
5.5 Transfer Functions 54
5.6 Impulse Response 56
5.7 Filter Cascades and Inverses 57
5.8 Chapter Summary 58
Exercises 59

6 Convolution and Related Theorems 61
6.1 Convolution Theorem for the Z Transform 61
6.2 DFT Circular Convolution Theorem 62
6.3 Autocorrelation Theorem 63
6.4 Window Functions 64
6.5 Linear Filtering with the DFT 68
6.6 Chapter Summary 70
Exercises 71

7 Least Squares 73
7.1 Motivations for Least Squares 73
7.2 Least Squares via Maximum Likelihood 74
7.3 Least Squares via Linear Algebra 77
7.4 Weighted Least Squares 81
7.5 Parameter Error Covariance Matrix 82
7.6 Fitting Data to Sinusoids 83
7.7 Ocean Tide Prediction 84
7.8 Seismic Tomography 86
7.9 A Model for Global Sea Level Change 88
7.10 Chapter Summary 91
Exercises 92

8 Linear Filter Design   94
   8.1   Introducing the *Z* Plane   94
   8.2   *Z* Plane Geometry – Stability and Invertibility   96
   8.3   Notch Filter Design Using *Z* Plane Geometry   98
   8.4   Differential Equation to Digital Filter Equation   100
   8.5   Derivative and Integration Filters   103
   8.6   Echo and Reverberation Filters   104
   8.7   Sampled Impulse Response Filter Coefficients   108
   8.8   Gravity Anomaly Calculations   109
   8.9   Ground Motion Amplification in an Earthquake   112
   8.10 Chapter Summary   114
   Exercises   114

9 Least Squares and Correlation Filters   116
   9.1   Least Squares Inverse Filters   116
   9.2   Yule–Walker Equations   117
   9.3   Interpolation Filters   120
   9.4   Prediction Error Filters   122
   9.5   Deconvolution Filters in Reflection Seismology   123
   9.6   Power Spectrum Estimate from the PEF   124
   9.7   Vibroseis and Matched (Correlation) Filtering   124
   9.8   Correlation Filtering in the Global Positioning System   127
   9.9   De-Blurring Filter Design   129
   9.10 Chapter Summary   133
   Exercises   133

10 Power and Coherence Spectra   136
   10.1 The DFT Periodogram   136
   10.2 Periodogram of White Noise   137
   10.3 Comparing Power Spectrum Estimation Methods   141
   10.4 Correlation and Coherence   145
   10.5 Coherence of Sea Level Variations   146
   10.6 Searching for Milankovitch Periods   147
   10.7 Chapter Summary   150
   Exercises   150

Appendix A   Matrices and Vectors   152
Appendix B   Fourier Transforms of Continuous Functions   155
Appendix C   Random Variable Concepts and Applications   169
Appendix D   Further Reading   188

Index   189

Preface

At the beginning of 2007, I started preparing notes for an upper division course at the University of Texas at Austin (UT Austin) for geophysics majors, surveying time series and related data processing topics and illustrating the importance of linear systems and filter concepts in geophysical problems including instrument design, seismology, and gravity. The goal was to prepare students for career opportunities, often in the geophysical exploration industry, or for further study in graduate school.

This textbook has been developed from the course notes that have evolved over the years, and its key features are:

- Providing the foundation for a survey course on time series and data analysis methods suitable for upper division undergraduate science majors and early career graduate students.
- Surveying both time series analysis (the sampling theorem, the Discrete Fourier Transform, linear digital filters, and spectral analysis) and least squares – an essential tool in the analysis of time series and many other data types and in digital filter design.
- Bringing together the comprehensive range of topics required for a survey course suitable for undergraduates, from material in an extensive literature of textbooks, monographs, and journal articles.
- Offering a balanced mix of time series and data analysis methods and example applications, some from exploration geophysics but many from other geophysical problems.

Courses similar to the one I teach at UT Austin are common in undergraduate curricula emphasizing exploration geophysics, where reflection seismology is a main application. Yet the same methods are used in other disciplines, including meteorology, astronomy, oceanography, and many engineering fields. Given this, and the availability of time series and other data types on the World Wide Web and elsewhere, this text can serve as a resource supporting these other disciplines in the form of both organized courses and research at graduate and undergraduate levels.

In an ideal world, students who take this course would have a background in linear algebra, probability, statistics, and perhaps Fourier transforms of continuous functions. However, in my years of teaching, I have found that even students who have been exposed to some of these foundational topics in earlier course-work might not have fully grasped them. The first three appendices at the end of the book aim to address this issue.

Instructors teaching undergraduate science majors, who are typically prepared with at most two years of university calculus, should cover Chapters 1, 2, and 3, in that order. This should motivate the topics and ensure a proper understanding of the foundational notions of the sampling theorem, statistics, decibels, complex numbers, sinusoids, and Fourier series, which are needed to appreciate linear digital filters and the Discrete Fourier Transform.

Instructors teaching students who already have a good understanding of the topics in Chapters 1 through 3 might survey them quickly and then spend more time on Chapters 4 through 7, covering the Discrete Fourier Transform, linear digital filters, and least squares.

Chapters 8, 9, and 10 focus on linear filter design methods, applications, and spectral analysis. Some of these topics, such as linear filter design, are sufficient for an entire semester course, so might be only partially covered in a first semester with an emphasis on specific examples. In that case, these chapters might be a starting point for a second semester course that includes further discussion of linear filter design methods, the Fourier transform of continuous functions, as presented in Appendix B, and the underlying probability ideas in Appendix C.

Appendix D provides a list of monographs and texts for further reading on many of the topics covered in this textbook, often at a more advanced level. I have proactively decided not to include references within the text because virtually all topics in the book are well established in the recent literature. As evidence of this, most modern computational environments such as MATLAB, Mathematica, python, R, and others include functions to implement virtually all the algorithms and methods surveyed here.

I am grateful to have undertaken graduate studies at the Scripps Institution of Oceanography. Important lessons on many of the topics in this book were learned from my advisor Richard Haubrich and other faculty members, including Robert L. Parker, and dissertation committee members George Backus and Walter Munk. I also appreciate having been a colleague of Milo Backus at the University of Texas, who helped me understand many aspects of time series analysis related to exploration geophysics.

Clark R. Wilson

1 An Introduction with Geophysical Time Series Examples

Spatial and temporal samples of physical quantities are among the most common data form in the geosciences and many other fields. Such data are called time series (even if they are samples along a profile in space). This chapter presents examples of geophysical time series in order to illustrate the sorts of scientific questions that may be addressed by the methods in Chapters 2 through 10. These time series are used in later chapters and exercises, and numerical values can be downloaded from www.cambridge.org/9781108931007. Virtually all the examples are available on the World Wide Web.

1.1 Global Mean Sea Level

Figure 1.1 shows samples of mean global sea level change every 10 days, observed by satellite altimetry (radar measurements of sea surface height). This and similar sea level time series are among the most studied in climate science. By eye you can tell that the rate of sea level rise is nearly 3 millimeters per year (mm/year) and using least squares, developed in Chapter 7, the rate estimated from the data is 3.4 mm/year, but this value would certainly change if it were estimated from just a portion of the time series. Chapter 7 shows that the least squares approach can be justified by both probability (maximum likelihood) and geometrical (linear algebra) arguments.

In Figure 1.1 the least-squares-fit trend line and the residual (detrended) data are both plotted by connecting the individual sample dots. The detrended sea level shows both a seasonal cycle and also variations over longer time scales, some lasting several years. The variability of the residual can be measured by variance or standard deviation (Chapter 2), and the way in which this variability is distributed over different time scales can be measured by the power spectrum (Chapter 10). Either least squares or the Discrete Fourier Transform (Chapters 3 and 4) could be used to find average seasonal variations, which are interesting because they reflect changes in both the amount of water in the oceans (exchanged with land in the hydrologic cycle), and ocean temperature causing thermal contraction and expansion. We might ask whether long period sea level variations are related to other climate signals (they are), using the correlation coefficient or spectrum of coherence to measure this (Chapter 10). The linear trend and seasonal cycle are elements of a predictive model for sea level change. Additional contributions to such a model could be developed using relationships with other climate signals, or via self-prediction methods (autoregressive prediction), both developed in Chapter 9 using least squares.

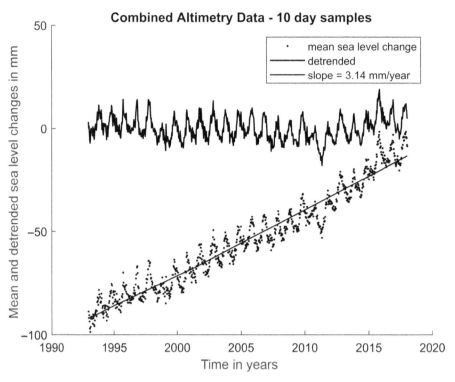

Figure 1.1 Global sea level time series, sampled every 10 days from combined satellite radar altimetry missions, courtesy of the University of Colorado Sea Level Research Group. Each 10-day sample is an average of many millions of individual radar travel time measurements, and a great effort has been made to calibrate such samples using tide gauges at the surface and to remove possible biases in the measurement. This is one of the most important time series supporting discussions of climate change. Developing a time series model to account for changes evident in this time series is key to addressing questions concerning human and other influences on the climate, and for purposes of predicting future sea level changes. One can see by eye that the average rate of sea level rise since 1992 is somewhat greater than 3 mm/year, but careful inspection shows that the apparent rate would differ if it were determined using only a subset of the entire series. For example, there are intervals where global sea level appears to rise more rapidly for several years, and others where it is steady or possibly declines. In Chapter 7, least squares is used to develop a model which partially explains some of these variations.

1.2 Stream Discharge

An important goal in applied hydrology is to develop methods of predicting stream flow following a rainfall event. Figure 1.2 shows a year-long time series of the daily stream discharge for Bull Creek, in Northwest Austin, Texas. The daily samples have been connected to clarify the repeated pattern, after a rainfall event, of very rapid rise in flow followed by a gradual decline over several days. Hydrologists call this pattern a stream hydrograph, and it represents the impulse response of the Bull Creek watershed to brief (impulsive) rainfall events. The plot shows that, approximately, the stream discharge time series is a superposition of many hydrographs, each scaled in amplitude by the amount of rainfall and shifted to the time of the rainfall event. This addition of scaled and shifted impulse responses is called convolution.

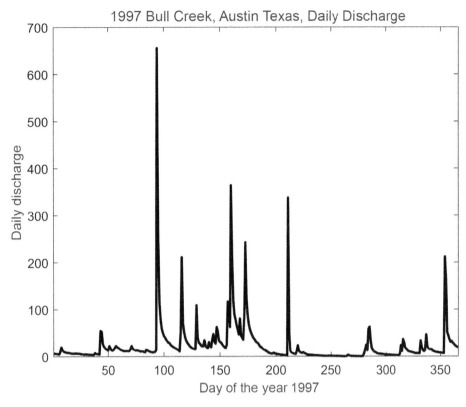

Figure 1.2 An important goal in hydrology is to predict the stream flow following a rainfall event. This figure shows, as an example, a daily time series of the stream flow for Bull Creek, Austin, Texas for the year 1997. Data are provided by the US Geological Survey. The time series shows the daily discharge in cubic feet per second. Daily samples have been connected in the plot to clarify the repeated pattern appearing with successive rainfall events. This pattern is the stream hydrograph. Each rainfall event initiates a rapid flow increase, with a slow decline over several days afterwards. Each event appears to add to the stream flow time series a stream hydrograph shape that is scaled in magnitude by the magnitude of the event. This is a linear filter model in which the stream hydrograph is the impulse response and the filter output is the stream flow, that is, the convolution of the stream hydrograph with the input rainfall time series. Further details of linear filters are given in Chapter 5. A linear filter model provides a simple way to predict stream flow from observed rainfall, but it is only an approximate method. It ignores the prior state of the watershed, which will affect runoff and stream flow. Conditions such as soil saturation and vegetation cover are known to be important. As computational power has increased over time, numerical stream flow predictions taking into account the changing state of the watershed have replaced the use of a linear filter approximation.

What we have described is a linear filter model, for which an input rainfall time series produces an output stream discharge time series equal (approximately) to the impulse response stream hydrograph convolved with the input. This time series provides an excellent visual explanation of convolution, a fundamental element of linear filter models that has been widely adopted to approximate often complex geophysical processes. However, more precise methods for stream discharge prediction have come into use with the advent of increasing computer power. A linear filter model ignores the prior state of the watershed which will affect runoff and stream flow. Changing conditions such as soil saturation and vegetation cover cause the impulse response of the watershed to vary over time.

Linear filters and convolution are discussed in Chapters 5, 6, 8, and 9. Chapter 5 reviews digital linear filter descriptions, including filter equations, the impulse response, the transfer function, filter cascades, and inverse filters. Chapter 6 deals with convolution and correlation and relationships with the Discrete Fourier Transform. Chapter 8 is a survey of various methods for designing linear filters for specific tasks, such as smoothing a time series, and for developing linear filter models of physical systems and processes, as in the Bull Creek example.

1.3   Eastern Pacific Sea Level

Tide gauges have been used for centuries to measure sea level variations at ports and harbors because these variations are of first-order importance in ship navigation. The tides, which change water levels by many meters at some locations, are the dominant cause of variations at periods shorter than one month. These are discussed in Chapter 7 in the context of tidal prediction. At periods longer than a month, other factors become important, including global climate change, which causes the steady rise shown in Figure 1.1. Superimposed on this steady rise are fluctuations due to a variety of causes. Measuring these intermediate time scale variations is important both to illuminate the underlying climate processes and for practical use in planning and designing harbor facilities. Figure 1.3 shows 29 years of monthly mean sea level change from tide gauges located on the west coast of North America and on the islands of Oahu (Honolulu) and Midway. The data are from the Permanent Service for Mean Sea Level (PMSL), an international organization collecting tide-gauge data from around the world; the data come from hundreds of sites and span decades to centuries. Data from the PMSL website are reported with an arbitrary mean value, so there is only information about sea level variability in these records.

For monthly averages, tidal variations (mainly the daily and twice daily variations described in Chapter 7) tend to average to zero. Long-period effects on tide-gauge elevation also contribute to apparent sea level change. For example, the record from Juneau, Alaska, shows a sea level decline over this period, possibly owing to the combined effects of local tectonics and elastic rebound and changing local geoid height as mass is lost from nearby melting glaciers. The rate of sea level rise indicated by Figure 1.1, slightly more than 3 mm per year, would produce a change of about 100 mm in 29 years. This would be barely perceptible in these plots in the presence of strong seasonal and other changes. One of the interesting features is the evidence of correlation between many stations at longer periods. In Chapter 10 we show how the variation of correlation with time scale may be measured using the coherence spectrum.

1.4   El Nino Southern Oscillation (ENSO) Index

Figure 1.4 shows one of a number of unitless indices used to judge the strength of the El Nino Southern Oscillation climate cycle. The series is shown starting in 1951 and is called the Multi-Variate ENSO Index (MEI); it is produced by the National Oceanic and Atmospheric Administration (NOAA) Physical Sciences Laboratory. Various ENSO indices, some extending back to 1871, have been computed for the purpose of understanding climate variations in the past. A portion of this

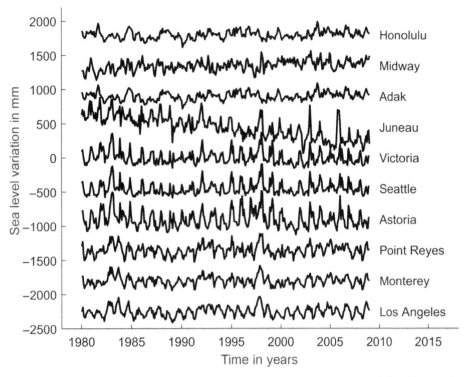

Figure 1.3   Time series from 8 tide-gauge stations on the west coast of North America and two in the central Pacific, from Midway Island and Hawaii. The series are arranged bottom to top in order of South to North location. They are taken from the monthly mean sea level recorded by the Permanent Service for Mean Sea Level (PMSL). The sea level shows seasonal and other variations. The seasonal variations are examined in Chapter 4 as an application of the Discrete Fourier Transform. By eye, one can see that several of the west-coast time series are correlated at periods longer than a year or two. These correlated changes are related to Pacific-wide climate processes, and an important goal is to identify these processes and eventually develop predictive models. The dependence of the correlation on the time scale of the variations can be quantified using the spectrum of coherence, as described in Chapter 10. A sea level rise at a rate of about 3 mm/year is evident in Figure 1.1 but is not apparent here owing to the relatively large seasonal and other variations. Long-period effects on tide-gauge elevation also contribute to an apparent sea level change. For example, the record from Juneau, Alaska, shows a sea level decline over this period, possibly owing to the combined effects of local tectonics, and the elastic rebound and changing local geoid height as mass is lost from nearby melting glaciers.

time series is included as one component of a least squares fit model to predict the global sea level variations in Figure 1.1. The development of this model is described in Chapter 7.

The original idea of the ENSO index was due to Gilbert Walker in the early twentieth century. Walker was investigating climate cycles that caused the Indian monsoon to vary in strength, and he developed an index based on barometric pressure differences between Darwin, Australia, and Tahiti. This is still regarded as the classical ENSO index; however, the MEI and other variants have been developed in recent years. Walker also developed prediction methods using autocorrelations, as will be described in Chapter 9, which lead to the Yule–Walker equations. These equations are the foundation of data processing in exploration geophysics, so there is a direct connection between Walker's climate science and the science of seismic reflection data processing.

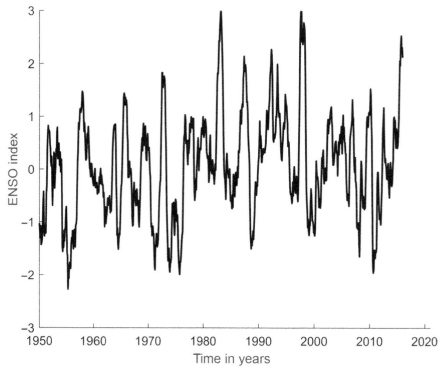

Figure 1.4 The Multi-Variate ENSO Index (MEI) produced by the NOAA Physical Sciences Laboratory. This is one of a number of indices designed to measure and predict the El Nino Southern Oscillation (ENSO) climate cycle affecting the Pacific Ocean and Indian Oceans and elsewhere. Historically, the Southern Oscillation was discovered by Gilbert Walker in the early twentieth century, as part of his effort to predict the strength of the Asian monsoon. Walker developed a Southern Oscillation Index derived from barometric pressure differences between Darwin, Australia, and Tahiti. Later in the twentieth century, it was recognized that an eastern Pacific ocean warming, known as El Nino, was related to the Southern Oscillation. El Nino (Spanish for "the baby", referring to the baby Jesus) was recognized by Peruvian fishermen around Christmas time in some years, when the fish harvest declined greatly due to warming of the surface waters. Today, climate predictions use this and other variants of the ENSO index to make long-range forecasts of the climate in North America and elsewhere. In Chapter 7 it is shown that a useful model for global sea level variations (the detrended curve in Figure 1.1) includes variations proportional to the multivariate ENSO index.

1.5 Lake Vostok Ice Core Temperature History

Figure 1.5 shows estimated temperatures over nearly the past half million years derived from an Antarctic ice core taken at the Lake Vostok station. The original data are available at the National Oceanic and Atmospheric Administration (NOAA) National Center for Environmental Information. Estimated temperatures are derived from oxygen isotopic variations contained in the ice. The data have been interpolated using a cubic spline method to a uniform sample interval of 500 years, to facilitate analysis by standard time series methods. A repeated cycle of ice ages is evident, the last ending about 20,000 years ago.

One of the great climate questions of the past several decades has been to identify the physical processes causing the regular cycle of ice ages evident in Figure 1.5. An early twentieth century theory by Milankovitch postulated that periodic changes in Earth's orbital characteristics would

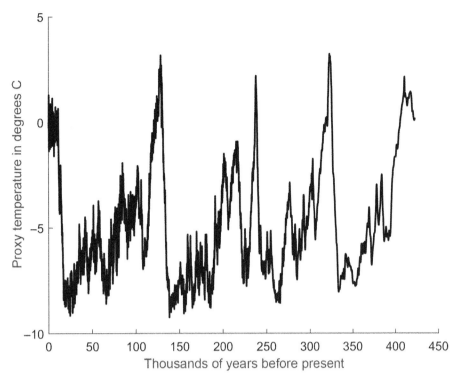

Time series of proxy temperatures estimated from the ice core at Lake Vostok Station, Antarctica, using oxygen isotope ratios. Time increases towards the left. The end of the last ice age began about 20,000 years before the present. Repeated ice ages every 100,000 years are evident. A period of 100,000 years is one of the Milankovitch periods associated with changes in Earth's orbital parameters: beginning in 1920, the Serbian geophysicist Milutin Milankovitch developed orbital variation theories that led to predicted periodic climate variations. Additional Milankovitch periods are at 26,000 and 41,000 years. Although well grounded in physical principles, his predictions were not widely accepted before his death in 1958. About two decades later, the first samples from deep ocean cores revealed Milankovitch-period climate variations. Subsequent sediment and ice core records continue to confirm their influence on the past climate. Here, samples taken from the ice core were interpolated to uniform intervals of 100 years using a cubic spline method to facilitate spectral analysis in a search for Milankovitch periods, as discussed in Chapter 10.

create thermal forcing mechanisms at several different periods, ranging from tens of thousands to a hundred thousand years. Milankovitch used astronomical observations and calculations and had no knowledge of past climate. Analysis of ice core proxy temperature records in terms of frequency content, via the power spectrum, is used to test whether Milankovitch periods are actually present in the past climate. This question serves as an exercise in Chapter 10.

1.6 Hector Mines Earthquake Seismograms

Figure 1.6 shows 960 seconds of seismic waves observed at the station PAYG in the Galapagos Islands due to the October 16, 1999, magnitude 7.1 strike–slip earthquake near the Hector Mines quarry in southern California. The data are available from the IRIS (Incorporated Research

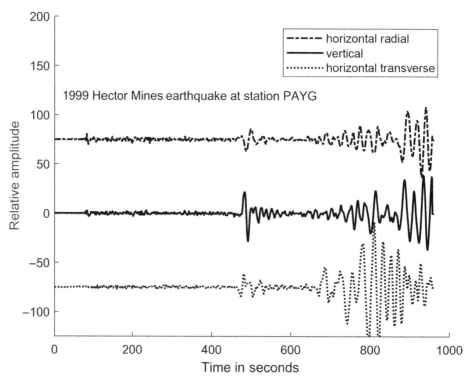

Figure 1.6 Seismograms from the 1999 Hector Mines earthquake in southern California as observed at the station PAYG in the Galapagos Islands. The seismometers at PAYG consist of vertical, north–south, and east–west horizontal broadband instruments. The two horizontal components have been combined to obtain horizontal components that are transverse and radial (in line with the direction of arriving waves). The P wave arrivals appear about 60 seconds after the beginning of these time series. The transverse horizontal component shows relatively small P wave arrivals, but strong Love waves arriving around 700 seconds. The Love waves are dispersed, with the longer-period waves arriving first. The vertical and radial components show strong Rayleigh wave arrivals at around the same time. The dispersion of both Love and Rayleigh surface waves is controlled by the crust and upper mantle elastic and density variations. The measured dispersion can be used to estimate these variations along the travel path between source and receiver, which in this case is largely oceanic.

Institutions for Seismology) data base. The horizontal components of the broadband recording have been combined to form separate seismograms for directions that are transverse and radial (in line with the direction of the arriving waves). The original seismograms give the north–south and east–west components.

The records begin about 60 seconds prior to the expected arrival time of the P wave. The transverse component is dominated by Love waves (transverse horizontal shear surface waves), which are dispersed, with the longer-period waves arriving earlier than the shorter-period waves. The P wave arrivals in the transverse component are much smaller than those in the vertical and radial components. A dispersed Rayleigh wave appears on both the vertical and radial components. Quantifying the dispersion provides a useful measure of crust and upper mantle seismic velocities along the travel path, which in this case is mainly oceanic crust. Chapters 3 and 4 show how the concept of the analytical signal, and the associated instantaneous frequency, can be computed using the Discrete Fourier Transform and used to quantify dispersion.

1.7 Simulated Seismograms, White Noise, and Computing Environments

Applications of linear filtering to exploration seismology (especially reflection seismology) appear in Chapters 7, 8, and 9 and in Appendix C. Simulated reflection seismograms in Chapter 9 and Appendix C were generated in the MATLAB computing environment to illustrate important concepts. A reflection seismogram is modeled as the convolution of a seismic wave impulse response with a sequence of reflection coefficients associated with physical properties varying in a vertical direction beneath the surface. This model and associated data processing methods based on least squares principles (see Chapter 7) are discussed in Chapter 9. The reflection seismogram model is similar to the linear filter model of stream discharge in Figure 1.2 as the convolution of an impulse response (the unit hydrograph) with a sequence of rainfall events.

A number of time series, either in the text or in the exercises, were created using computer random number generators. These are actually pseudo-random number generators, which are algorithms that would reproduce exactly the same sequence of numbers without changing an algorithm parameter known as a seed. However, pseudo-random sequences of numbers do behave as if they were random, that is, independent from one value to the next. A time series of such numbers is called white noise. White noise properties will be described in Chapter 2. The type of random numbers (their distribution) is often Gaussian or normal, as explained in Appendix C. Pseudo-random number generators are used to create a random binary sequence (values are 0 or 1) as the key element of the Global Positioning System signal, as explained in Chapter 9.

The MATLAB computing environment was used to generate many of the figures throughout the text and appendices, but other environments or languages (python, Mathematica, R, Octave, etc.) would also be suitable. For all these languages and environments, specialized functions are available to implement digital filtering, filter design, spectral analysis, and other techniques. As the minimum, one needs a Fast Fourier Transform algorithm (Chapter 4); software to implement a linear digital filter (Chapter 5) (writing a convolution algorithm is presented as a Chapter 5 exercise); a linear equation solver (Chapter 7); and a random number generator (Chapter 2). Only a random number generator for a uniform distribution is strictly required. A Chapter 2 exercise shows how to use it to create approximately Gaussian-distributed random numbers. The sum of squared Gaussian random numbers creates a chi-squared distribution, so all three of the important random variable distributions (Appendix C) can be generated with a uniform generator.

1.8 Chapter Summary

The geophysical time series examples in this chapter serve to motivate and illustrate important data processing concepts and methods developed in later chapters.

- The global mean sea level time series (from satellite radar altimetry, illustrated in the cover photograph) is widely recognized as an important measure of contemporary climate change. It provides a useful case study in the least squares development and assessment of time series models in Chapter 7.
- The stream discharge time series provides an excellent visual illustration of convolution, a central concept in linear filters, developed in Chapter 5.

- The suite of eastern Pacific sea level time series gives a perspective on sea level variation distinctly different from that of the satellite radar altimetry series. At individual ports and harbors local effects dominate, while the average global rise in the altimetry series is obscured. Chapter 4 will show how the Discrete Fourier Transform can be used to find seasonal variations in these sea level time series, and Chapter 10 will demonstrate how the spectrum of coherence is useful in separating spatial scales of forcing mechanisms.
- The ENSO index series is an important measure of contemporary climate change. It is used in Chapter 7 as an element of the global sea level model developed using least squares.
- The Lake Vostok ice core temperature record appears in Chapter 10 as an example of the use of spectral analysis to look for the periodic climate forcing proposed by Milankovitch in the twentieth century.
- Seismology is a mainstay of geophysical investigations at many different spatial scales. The Hector Mines earthquake seismograms provide a global-scale example. They are used in Chapter 4 in an example application of the Discrete Fourier Transform to quantify the dispersion of seismic surface waves. Applications to seismic methods at smaller exploration scales appear in Chapters 7, 8, and 9.

Analog Signals and Digital Time Series

In this chapter we describe the motivation and requirements for converting analog signals to digital time series so that they can be analyzed using digital computers. The process of digitizing analog geophysical signals, for example continuous voltages from seismometers, must ensure that all information about signal amplitudes (the dynamic range) and range of frequencies (the frequency bandwidth) is retained. Important elements of analog to digital conversion include the number of samples to be taken (the sampling rate) and how to represent samples as numbers in a computer. The chapter also presents the standard statistics used to describe time series and introduces the logarithmic decibel scale as a convenient way to compare the relative magnitudes of signals.

2.1 Digital Time Series Notation

We use an expression like $s(t)$ to represent a continuous (analog) function of time t. The lower case letter s is chosen to remind us of the word seismogram, if that is the type of signal. The function $s(t)$ is like those encountered in a calculus class in that it is continuous. By convention, an upper case letter (for example S) is used for the frequency domain description of the time series designated by the corresponding lower case letter. A digital time series containing samples of $s(t)$ is an ordered set of numbers which can be stored in a computer memory. The details of how these samples relate to the analog signal depend on the electrical device (analog to digital converter or ADC) used to do the conversion. The properties of ADCs vary, so here we will assume that an ADC measures a time-dependent voltage $s(t)$ and creates a set of samples that are instantaneous values of $s(t)$ at a uniform time increment Δt. The sampling frequency is $f_s = 1/\Delta t$, measured in hertz (Hz) when t is time in seconds.

The sampled time series is denoted by

$$s_t = [s_0, s_1, \ldots]$$

We use the same lower case letter to indicate the time series, but the meaning of the subscript t is different. The values are now integers, so here t is the sample number. Since the digital sample interval Δt is known, converting sample number to clock time is easy. If $s(t)$ is of infinite length then s_t is likewise infinite, but this is only of theoretical interest because, in all practical situations, a time series must be of finite length to fit within computer memory. Usually, zero time is taken as the start of a time series, so the first sample is s_0. The expression s_t may refer to an entire time series, or just to the value of that series at a particular time t. The context is usually clear. If s_t is a particular value at time t, then s_{t+1} is the sample just after that and s_{t-1} is the sample just before.

If s_t refers to the entire time series with N samples, then $s_t = [s_0, s_1, s_2, \ldots, s_{N-1}]$ is an ordered array of N numbers. This is an N-vector (vector of length N), which might be written either as a row vector (as shown) or as a column vector, depending on what is needed. We may also use a simple lower case s in place of s_t to refer to the vector. The one-to-one correspondence between time series and vectors leads to many important applications of linear algebra in data processing.

2.2 Digitizing Analog Signals

The sampling theorem describes the requirements for digitizing continuous analog signals (typically the time varying voltage from an instrument) at a uniform rate, and the penalty (aliasing – confusion of frequencies) for taking insufficient samples. For electrical or similar signals, the requirements of the sampling theorem and avoiding aliasing are the focus of this section. The basic strategy is to smooth signals prior to sampling. However, aliasing cannot always be controlled in geophysical time series. An example is a time series of repeated measurements from a satellite flying over a point on Earth. The time between samples may be irregular, depending on the satellite orbit characteristics, and may be many days apart, but the measured quantity may be continuously and rapidly varying between samples. This makes aliasing a likely problem, and one that is difficult to understand and correct. Removing aliased signals (de-aliasing) may require the construction of models to predict rapid variations contributing to the sampled data so that they can be subtracted.

The digitizing or sampling task has the goal of taking an analog signal, often a voltage varying in time, and turning it into a set of discrete samples, consisting of numbers, that will retain all information in the analog signal. Another issue is how to format the numbers (how many bits and how they are used) in a way that allows the full range of signal amplitudes to be recorded.

We will assume that sampling is uniform, with samples taken at a constant rate. The sampling frequency f_s is the reciprocal of the time interval between samples, Δt, so $f_s = 1/\Delta t$. The smoothness of the analog signal clearly determines how small Δt must be. A qualitative notion of smoothness can be judged from a plot of the signal as a function of time. A quantitative measure of smoothness is obtained by determining the highest sinusoidal frequency in the signal. The power spectral density (PSD) measures this, as will be described in Chapter 10.

The sampling theorem states the requirements for the sampling frequency in terms of the highest frequency contained in the analog signal. That an analog signal has a definite highest frequency is a statement that $s(t)$ is band-limited. That is, its power spectrum is zero beyond some highest frequency. For geophysical signals obtained from instruments such as seismometers, signals are always band-limited. Of course, the actual measurements might have features that are not band-limited, such as a sudden offset in the voltage due to an electronic malfunction. Before or after digitizing, such problems must be somehow detected and corrected.

The sampling theorem applies to the situation where an analog signal is uniformly sampled. In its complete form (as the Nyquist–Shannon sampling theorem) it presents both sampling rate requirements and a proof that it is possible to reconstruct the original analog signal from the discrete samples. Here we simply state the theorem and omit the proof, outlined in Appendix B. The sampling theorem was developed in the early and mid twentieth century and is considered the foundation of modern digital signal theory. However, there have been efforts to overcome its requirements because it often leads to the recording of many samples (large data sets), which are cumbersome to store and transmit. This is well known in music recording, where smaller file sizes are obtained with formats such as MP-3. In other applications samples are not always available at a uniform rate, requiring specialized algorithms to deal with original irregular samples or assumptions that allow interpolation to a uniform rate.

The sampling theorem can be stated as follows. An analog signal may be reconstructed from its digital samples when it has been uniformly sampled at intervals of Δt, provided that it contains no frequencies greater than the Nyquist frequency $f_N = f_{Nyquist} = 1/(2\Delta t)$, which is half the sampling frequency.

An equivalent statement is: There must be at least two samples for every cycle of the highest frequency sinusoid present in the analog signal.

Another equivalent statement is: The sampling frequency ($f_s = f_{sampling} = 1/\Delta t$) must be at least twice the highest frequency in the analog signal.

The sampling theorem promises that you can reconstruct the analog signal if you obey this rule. The details of reconstruction in Appendix B show that analog signal reconstruction uses a linear combination of the discrete samples. The main point at present is to understand the sampling theorem requirements and to recognize problems that arise when these requirements are not met.

If the sampling frequency is chosen in advance of data collection, or constrained by the digitizing equipment, then it is essential to condition the analog signal before it is sampled. The conditioning required is to make it smoother, which means to remove frequencies above the Nyquist frequency $f_N = f_s/2$. Smoothing is done with a filter (usually analog) that rejects frequencies above the Nyquist frequency. Because analog filters are not perfect, in practice this requires cutting frequencies somewhat below the Nyquist, so that virtually no analog signal remains at the Nyquist frequency and above. An anti-alias filter is a low-pass filter because it passes low frequencies and rejects those higher than the specified Nyquist frequency. In some instruments the anti-alias filter involves a combination of analog and digital filters. For example, the ADC may sample at a rate much higher than the final sampling frequency in the recorded data. This is called oversampling and is followed by application of a digital low-pass anti-alias filter, which allows resampling at a lower rate, a process called decimation, to allow the data to be recorded.

The anti-alias filtering of electrical analog signals (voltages) is straightforward and virtually all instruments that record digitally employ a proper anti-alias filter. However, it is not always possible in other situations. For example, if the Weather Service records hourly samples of temperature, no smoothing is first applied to the temperature variations. Because temperatures might vary rapidly between hourly readings, there is likely to be some violation of the sampling theorem.

Three scenarios are possible when converting analog to digital signals. The first is oversampling: acquiring more samples (using a smaller Δt) than necessary. Oversampling means that the Nyquist frequency (half the sampling frequency) greatly exceeds the highest frequency in the analog signal. This does no harm, but could create a burden in handling and storing unnecessary data. However, some amount of oversampling is useful and common. Oversampling can be a deliberate part of the anti-alias filtering process when it is followed by the application of a digital filter that smooths the data prior to resampling it at a lower rate (decimation), in order to achieve the final desired sampling rate. When there are just enough samples, called critical sampling, the highest frequency present in the analog signal is exactly equal to the Nyquist frequency. This can be dangerous because it is close to the third condition which is undersampling.

2.3 Undersampling and Aliasing

Undersampling violates the sampling theorem, so the analog signal cannot be recovered from its samples. Undersampling not only loses information about the analog signal at frequencies above the Nyquist but also contaminates frequencies below the Nyquist. This contamination in the Nyquist band is called aliasing. An alias is a false name and in this context means that the high frequencies (above the Nyquist) have precisely the same samples as frequencies within the Nyquist band. They are thus indistinguishable from those within the Nyquist band, and contaminate them. Making the signal $s(t)$ smooth before sampling, by applying a low-pass (or high-cut) anti-alias filter, solves the problem.

It is common in geophysical data analysis to change the sample interval Δt of a time series from its original value. Resampling to a smaller Δt (taking more samples) is called interpolation, whereas resampling to get fewer samples is called decimation. There are many methods of interpolation that work with both uniformly sampled and irregularly sampled data. The general idea is to find a smooth curve that passes through all the original data. This curve can be sampled at whatever times are

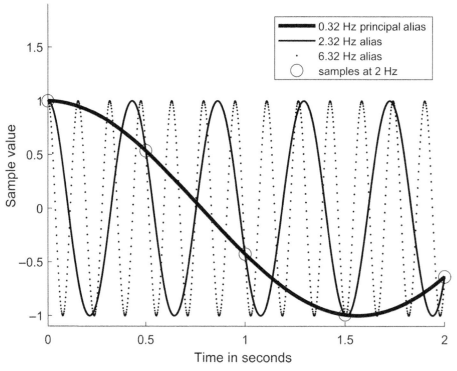

Figure 2.1 A sinusoid with frequency 2.32 Hz (lighter solid line) is sampled every half second (therefore, the sampling frequency is $f_s = 2$ Hz). Because 2.32 Hz is above the Nyquist frequency of 1 Hz (half the sampling frequency), the samples could correspond to those of the principal alias at 0.32 Hz. Other alias frequencies would be 0.32 Hz plus higher integer multiples of f_s. For example, 4.32, 6.32, and 8.32 Hz sinusoids would all have the same samples. In fact there is an infinite number of sinusoids at frequencies outside the Nyquist frequency band that would have exactly the same samples if these were taken every half second. An anti-alias filter is used to condition a time series so as to remove all frequencies outside the Nyquist band. In this way, samples of the principal alias are not contaminated by samples of its aliases outside the Nyquist band of frequencies.

desired. The corresponding Nyquist frequency will be higher than the original, so in the ideal case the interpolation will not contain any higher frequencies (above the old Nyquist value). We will see later that it is possible to do this ideal interpolation using the Discrete Fourier Transform. Other interpolation methods such as linear or cubic spline, which use polynomials, are popular and efficient but will add frequencies above the old Nyquist frequency in an unpredictable way.

Decimation also requires some care because it means that, with fewer samples, the new Nyquist frequency is lower than the original. This requires removing variations at frequencies between the two Nyquists, usually by smoothing (low-pass filtering) prior to resampling. This smoothing is done with a digital filter, because the data consist of a digital time series.

The fundamental cause of aliasing is that there are an infinite number of sinusoids of different frequencies, all of which could correspond to the same samples. Figure 2.1 shows an example in which three different cosine functions have the same samples, making them, by definition, aliases of one another. The associated frequencies are at 0.32 Hz, the principal alias, and two frequencies above the Nyquist, separated from the principal alias by integer multiples of 2 Hz, the sampling frequency. Figure 2.1 leads us to conclude that all aliases of frequency f within the Nyquist frequency band are separated from it by integer multiples of the sampling frequency. That is, all alias frequencies are separated by $n\Delta t$, where n is an integer. In Chapter 3 we will see that n may be either a positive or negative integer, requiring that we allow for both positive and negative frequencies. The distinction between positive and negative frequencies will become clear in the context of complex sinusoids, developed in Chapter 3.

2.4 Time Series Statistics

A statistic is a number derived from a data set; the use of such statistics is motivated by the need to summarize a large number of data in a succinct way. For example, in the financial industry, various statistics (the Dow Jones Industrial Average and others) summarize the status of a large and complex stock market. For a time series x_t of length N, three simple statistics, the mean, variance, and standard deviation, are in common use. The mean gives the central value of the time series, while the variance indicates the average squared departure from the mean. Its square root, the standard deviation, measures variability in the same units as those of the time series. A fourth statistic, the autocorrelation, is a sequence of numbers. The autocorrelation statistic contains information about the time series smoothness, as does the frequency content (spectrum) of the time series. Computation of these four statistics carries the underlying assumption that the time series x_t is stationary, that is, its statistical properties are unchanged over its duration. Stationary time series are our focus throughout.

If x_t is a portion of a stationary time series of infinite length, or an example from an infinite number of similar time series, then conceptually there are true values of these statistics that would be approached as infinite numbers of data are collected. So, we can view statistics either as numbers computed from the particular time series at hand or alternatively as estimates that would improve as more data became available.

2.4.1 Mean or Average Value

The sample mean of a time series, identified by the symbol $\hat{\mu}$, is the average value and is computed by adding up all time series samples and dividing by the total number of samples. A subscript may be added to the symbol for this and other statistics to identify the particular time series. We reserve μ, without the circumflex, to indicate a true mean value, which, in conceptual terms, is the limiting value of $\hat{\mu}$ obtained if the time series length N were infinite. The circumflex denotes $\hat{\mu}$ as an estimate of μ. The actual computation of the mean value is given by the summation

$$\hat{\mu} = \frac{1}{N} \sum_{t=0}^{N-1} x_t$$

where x_t is the signal at time t and N is the number of samples or time intervals. The mean has the same units as the time series. Depending on the physical problem, it may or may not be interesting. For example, if the time series consists of hourly values of air temperature, then the mean temperature is an important climate measure. On the other hand, the mean voltage from a seismometer is often near zero, and usually of no interest, because the interesting quantity is the variation about the mean as a seismic wave passes by. In this case, it would be common practice to compute the mean and subtract it from each value, leaving a zero-mean residual time series for further analysis.

In some cases another statistic, the median value, is used to describe the central value of a time series. The median is the middle value, with equal numbers of larger and smaller samples. Finding the median normally requires the time series to be ordered ("sorted").

2.4.2 Variance and Standard Deviation

The variance $\hat{\sigma}^2$ is the average squared deviation of the time series from its mean. The variance is associated with the concept of power, the rate of energy delivery because, in many cases, energy is proportional to squared amplitude. For example, in a stretched spring, the stored energy is proportional to the square of the spring displacement, so the variance of the displacement time series would measure the average stored energy per unit time, or average power. The variance is also an important statistic used to judge how well a model fits a time series. It is central to least squares methods, which minimize the variance of the misfit between a time series and a model.

To compute the variance $\hat{\sigma}^2$ is computed by first subtracting the mean from each value of the time series, leaving a zero-mean residual. The terminology "sample variance" clarifies that $\hat{\sigma}^2$ is determined from a particular set of data samples while σ^2 is used to refer to the true variance. As with the mean, the sample variance can be taken as an estimate of the true variance. The actual computation formula can be either

$$\hat{\sigma}^2 = \frac{1}{N} \sum_{t=0}^{N-1} (x_t - \hat{\mu})^2$$

or

$$\hat{\sigma}^2 = \frac{1}{N-1} \sum_{t=0}^{N-1} (x_t - \hat{\mu})^2$$

When the true mean is unknown, the scaling $1/N$ is commonly replaced by $1/(N-1)$, to make $\hat{\sigma}^2$ an unbiased estimate of the true variance σ^2. This is discussed in Appendix C.

The standard deviation, also called the root mean square or rms value, is $\hat{\sigma}$, the square root of the variance. The information contained in either the variance or standard deviation is the same, but the standard deviation has the same units as the time series x_t so may be easier to interpret.

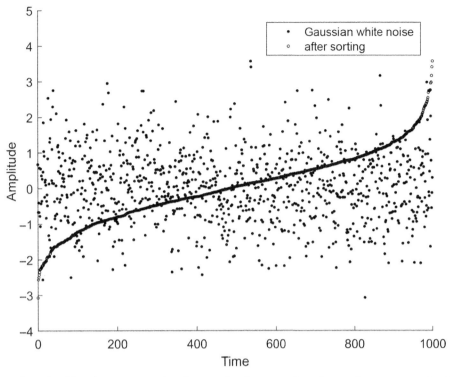

Figure 2.2 One thousand samples of Gaussian zero-mean unit-variance numbers from a pseudo-random number generator create a time series of Gaussian white noise. If the white noise series is sorted (solid line formed by open circles) then the time series is no longer white noise but is dominated by low-frequency variations, and every value in the sorted series is related to every other.

2.4.3 Autocorrelation

The autocorrelation of a time series is a sequence of numbers computed from sums of products of time series values separated (lagged) by an integer number of samples, called the lag, τ. As an example, the autocorrelation of the time series $h_t = [h_0, h_1, h_2]$ is

$$r_\tau = [h_0 h_2, h_0 h_1 + h_1 h_2, h_0^2 + h_1^2 + h_2^2, h_0 h_1 + h_1 h_2, h_0 h_2] = [r_{-2}, r_{-1}, r_0, r_1, r_2]$$

The central value, at lag zero (r_0), is the sum of the squares of the time series values, equal to the variance if we had divided by time series length (number of samples). The autocorrelation is symmetric about zero lag, so the same values appear at negative and positive lags (negative and positive lags are complex conjugates in the case of complex-valued time series). For real-valued time series, both h_t and its time-reversed version $h_{-t} = [h_2, h_1, h_0]$ share the same autocorrelation. The lag 1 or -1 values are the sums of products of adjacent time series values. The lag 2 or -2 values are the sums of products of values separated by two time samples, and so on. In the data processing literature, the autocorrelation is normalized in various ways, for example, by dividing all values by the number of samples or the lag zero value. In a later chapter we introduce the pentagram notation for correlation, in which the autocorrelation of a time series h_t is $r_\tau = h_t \star h_t$.

   If the autocorrelation is considered as a statistic of a time series then, like the mean, variance, and standard deviation, there is an estimated value computed from data in hand, and conceptually also a true value obtained as the number of data grows large. To be consistent with notation for the other statistics, we might use a circumflex to indicate the estimate, as $\hat{r}_\tau$. This is not common in the literature, and is not used here, so r_τ may be either the estimated or true autocorrelation, depending on the context.

The autocorrelation statistic plays a central role in the development of prediction, interpolation, and related filters in Chapter 9, while Chapter 6 shows that it contains the same information as the periodogram power spectrum developed in Chapter 10. So, while applications will come in later chapters, here we develop a qualitative understanding of autocorrelation using a simple example.

Figure 2.2 shows two time series, both 1000 samples in length. One is a sequence of zero-mean random numbers, obtained using a computer algorithm known as a pseudo-random number generator. The other is the same set of numbers after sorting by increasing value, resulting in a smooth time series. The unsorted random numbers constitute a white noise series. Chapter 10 explains that the term "white" implies that the time series power spectrum contains all frequencies in equal amounts, in analogy with white light.

Because the same numbers are contained in both series in Figure 2.2, their means, variances, and standard deviations are identical, and therefore cannot be used to distinguish the two very different series. On the other hand, their autocorrelations, in Figure 2.3, are quite distinct. White noise autocorrelation values are small, except at zero lag, while the sorted series autocorrelation changes slowly with lag. This distinctive white noise autocorrelation behavior (nearly zero except at zero lag)

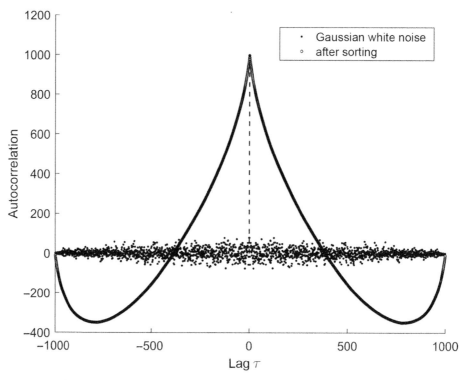

Figure 2.3 The autocorrelations of the white noise and of the sorted series from Figure 2.2 are distinctly different. The same set of 1000 numbers is present, so that the mean, variance, and standard deviations are identical for both the white noise and sorted series. The autocorrelation of the sorted series (bold curve, formed by adjacent dots) is smooth, and changes slowly with lag. The autocorrelation of the white noise series (black dots) is nearly zero except at zero lag, where its value (equal to the sum of squared values) is the same as that of the sorted series at zero lag; its distinctive behavior (nearly zero except at zero lag) provides a useful way to precisely align time series with white-noise-like autocorrelations.

has enabled many innovations in geophysics and other fields. Chapter 9 describes applications to the Global Positioning System, and to passive seismic methods.

2.5 Numerical Representation of Samples

The sampling theorem states the number of samples required to reconstruct an analog signal but not how to represent them as numbers in a computer, where a fixed number of binary bits is available for each sample. Thirty-two is a common number of bits, often associated with the term single-precision, so 64 bits corresponds to double-precision, 128 bits to quadruple-precision, and so on. One bit is the simplest binary (base-2) number, with a value of either 0 or 1. A set of eight bits is a byte. The numbers in a computer memory must record both the range of signal amplitudes, smallest to largest (the dynamic range), and the smallest change in amplitudes of interest (the precision). Usually the dynamic range is expressed in decibels, discussed later in this chapter. The two choices for making use of the finite number of bits are either integer or floating point formats. These correspond to the two numerical variable types, integer and real, in most computer languages. The integer format provides excellent precision but limited dynamic range, while the floating point format provides excellent dynamic range but limited precision. In computation, integer variables are commonly used to count the number of steps in a repeated operation or to identify the position of a value in an array of numbers, while floating point variables usually represent data values. The original samples from the ADC may be in either format. Computer languages typically have functions to convert integer to floating point or floating point to integer, and a language compiler will usually automatically perform a conversion, when the two are used in the same arithmetic expression, if the conversion is not done explicitly.

A simple example will illustrate the difference between the floating point and integer representations. Assume that only four bits are available to record the voltage output from a seismometer. A four-bit integer number may use one bit for the sign, leaving three for the signal amplitude, or all four may be used for a positive number. Let [xxxx] represent a four-bit binary number, where x can be either 0 or 1, and the first bit is for the sign. Thus if a negative number is taken to be [0xxx] then the corresponding positive number is [1xxx]. The remaining bits record powers of two: $2^2, 2^1, 2^0$. Thus a four-bit binary integer ranges over $[0111], \ldots, [0001], [1000], [1001], \ldots, [1111]$, corresponding to the 15 decimal numbers $[-7, -6, -5, -4, -3, -2, -1, 0, 1, 2, 3, 4, 5, 6, 7]$ or a total voltage range of 15, which is $2^{(number\ of\ bits)} - 1$. Negative zero was omitted from the list, though it is a possible number in this scheme. A change in value by 1 can be represented over the entire range, so precision is constant over the full range of numbers. If N bits are used in an integer format, the dynamic range is the square of $2^N - 1$. The dynamic range is often expressed in decibels, which will be discussed shortly.

The alternative to the integer format is to use the four bits to represent floating point numbers, the equivalent of scientific notation. A floating point number is represented as the product of a significand m (also known as a mantissa) and a base integer b raised to an exponent c. To store the number $m \times b^c$ in computer memory, the available four bits are used to store the sign and the two numbers m and c. With three bits available for m and c, suppose we use two of them for m and one for c, and set the base $b = 10$. With this allocation of bits, the exponent can take on the values $[0, 1]$ and m can have the values of $[0, 1, 2, 3]$, so the range of numbers possible is $[-30, -20, -10, -3, -2, -1, 0, 1, 2, 3, 10, 20, 30]$. Again, negative zero has been omitted, as well as positive or negative 0×10^1. Now the dynamic range is much larger than for the integer format, but with a loss of precision.

Small differences between large numbers cannot be represented. For example, if we subtract a small number (for example 1) from a large number (for example 30), the exact answer (29) cannot be represented. The result would be given as the closest value (30) and therefore incorrect owing to the lack of numerical precision. This problem exists with all floating point arithmetic regardless of the number of bits used, and is illustrated in some computations in the exercises. Using more bits can improve the numerical precision, so double (64-bit) or quadruple (128-bit) precision arithmetic may be used if available. Additionally, problems due to a lack of numerical precision can be reduced by conditioning (scaling or normalizing) quantities to avoid arithmetic operations with numbers of very different sizes.

2.6 Decibels

It is standard practice to measure one variance relative to another using the logarithmic decibel (dB) scale. The dB scale is ubiquitous in acoustics and electrical engineering, and was an invention of electrical engineers. The name is taken from the Latin *deci*, meaning factor of 10, and bel, a short form of Alexander Graham Bell, inventor of the telephone. When a value is given in decibels, there is always an underlying reference value. For example, when expressing dynamic range in decibels, the reference value is the square of the smallest non-zero signal amplitude that can be recorded or represented.

The decibel scale in acoustics is used to describe the large range of pressures sensed by the human ear. By convention, the reference pressure is the quietest sound that a human can hear, a whisper in a quiet room at a pressure of 20 micropascals and a frequency of 1 kHz. In this case the reference value is an amplitude rather than a variance. In general decibels may be calculated using either a reference variance or amplitude as follows:

$$\text{decibels} = 10 \times \log_{10}\left[\frac{signal\ \ variance}{reference\ \ variance}\right] = 20 \times \log_{10}\left[\frac{signal\ \ amplitude}{reference\ \ amplitude}\right]$$

The decibel scale is commonly used to plot the vertical coordinate in the plot of a power spectrum or other positive quantity with a large range of numerical values. The logarithmic scale compresses the vertical values, allowing the presentation of large and small values on the same graph. The dB scale is also the standard way to describe amplification, so it is useful to be familiar with a few key values. When a signal is doubled in amplitude, the amplification (the gain) is

$$20 \times \log_{10}(2) = 6.02 \approx +6 \text{ dB}$$

The reference value in this case is the signal amplitude before amplification. Similarly, reducing the signal amplitude by one-half gives a change of −6 dB, so negative decibel values imply attenuation (reduction in amplitude) and positive decibels mean amplification. A value of 0 dB means that a signal is the same size as the reference (the logarithm of 1 is zero). If a signal is doubled in amplitude, and then doubled again in amplitude, the final signal has four times the original amplitude, or an increase of +12 dB, with +6 dB produced at each doubling, so multiple steps of amplification in succession (a filter cascade) just add up the decibel effect of each.

A doubling of amplitude is a change of +6 dB, but if the amplitude is increased by a factor 1.414 (the square root of 2) then the power is doubled, because power is proportional to the square of amplitude. Therefore doubling the power corresponds to an increase of +3 dB. Experiments with human hearing show that 3 dB changes in sound level are perceptible to most people, so 3 dB increments are commonly indicated on the volume control knobs of sound amplifiers. The half-power point (-3 dB point) is also the standard way to specify the frequency at which a filter starts to cut or remove frequencies. The cutoff frequency is by convention that where the filter gain is 3 dB or more below its maximum.

2.7 Applications to Digital Audio Recording

Digital audio recordings became standardized with the development of the compact disc (CD) in the late 1970s. The technical specifications adopted for the CD can be understood using many of the concepts reviewed in this chapter. The CD was designed to provide a digital format well matched to human hearing, both in dynamic range and in frequency bandwidth. The only other major design parameter for the CD format was its size, set to match the length of the ninth symphony of Beethoven.

The dynamic range of human hearing is about 100 dB. At a sound level of 100 dB, the pressure on the eardrum is therefore 100,000 times that of the reference pressure, which corresponds to the

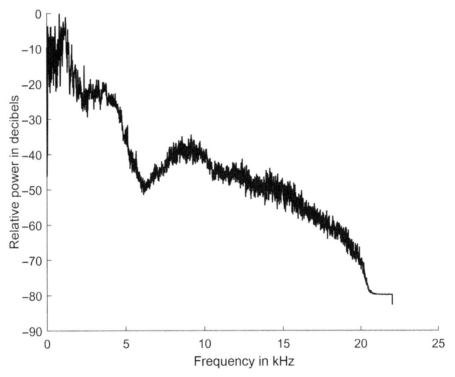

Figure 2.4   The power spectrum of about 38 seconds of a CD audio recording digitized from analog magnetic tape. The spectrum has been computed using a periodogram method. As described in Chapter 10, periodogram estimates typically show irregular oscillations, which in this figure make the curve appear relatively wide. The frequency range is [0, 22.05] kHz. The vertical scale in decibels gives the power relative to the maximum value, so all dB values are negative. The recording has most of its power below 5 kHz. An anti-alias filter was applied to the magnetic tape analog signal prior to digitizing at the CD sampling frequency of 44.1 kHz. The effect of the anti-alias filter is evident for frequencies at 20 kHz and above, where power drops sharply. The anti-alias filter reduces signals to a very small level at the Nyquist frequency of 22.05 kHz, so aliasing is imperceptable in the digital recording.

quietest sound that can be heard. The frequency bandwidth of human hearing is approximately within the range of 10,000 to 20,000 Hz, although it varies among individuals and usually changes with age. Sixteen-bit integers were chosen to represent digital samples in the CD format. These provide a dynamic range of 96 dB $\left(=10 \times \log_{10}\frac{(2^{16})^2}{1}\right)$. Here, the numerator in the logarithm is the square of the largest 16-bit integer (65,536), and the denominator is 1, the smallest positive integer number. By expressing the dynamic range in decibels it is easy to understand that the 16-bit integer format closely matches the 100 dB range of human hearing.

The CD sampling frequency was set at 44.1 kHz, making the Nyquist frequency of 22.05 kHz somewhat higher than the typical range of human hearing. Analog signals from microphones and recording media such as magnetic tape contain frequencies above this, requiring an anti-alias filter prior to sampling. Anti-alias filters are analog devices that do not provide a sharp frequency cutoff, so in practice they begin to attenuate signals somewhat below the 22.05 kHz Nyquist frequency. This can be seen in Figure 2.4, which shows the power spectrum of a CD recording digitized from an analog magnetic tape. The spectrum is computed from about 38 seconds of a female singer recorded on magnetic tape in 1956 and later transcribed to CD format. The effect of the anti-alias filter is evident above about 20 kHz, where the power level drops rapidly. The decibel scale used in the plot is convenient to display the large range of power spectrum values.

2.8 Chapter Summary

This chapter has surveyed analog signal sampling requirements for the case of uniform sampling, introduced various statistics as time series descriptors, and presented the decibel scale as a means of comparing signal strengths. The main points are as follows.

- Sampling a continuous (analog) signal in time or space requires at least two samples of the shortest-period or shortest-wavelength sinusoidal variation contained in the signal. An analog signal can be considered to be a superposition of sinusoids of many periods or wavelengths, as will be discussed in the next chapter.
- The penalty for taking too few samples (undersampling) is aliasing. Aliasing results in a confusion of frequencies in which many different sinusoidal frequencies yield exactly the same sample values.
- In some geophysical problems the sampling rate cannot be controlled. In such cases aliasing may be present but can possibly be reduced by developing models for the undersampled signals in a process of de-aliasing.
- The computer numbers used to represent data samples may be in either integer or floating point format. Numerical computations are usually done in floating point format, and numerical precision errors may arise in calculations, especially when both very large and small numbers are involved.
- The time series statistics mean, variance, and standard deviation provide useful summaries of a time series, giving the middle value (mean), and the variability about the middle value (variance or standard deviation). The autocorrelation derived from a time series provides a measure of how smooth it is. It contains the same information as the power spectrum, which describes how the time series variance is distributed over frequency.
- The logarithmic decibel scale is widely used to compare the relative sizes of signal variances and amplitudes. The decibel scale is the standard in plotting quantities such as the power spectrum and linear filter transfer functions, as will be developed in later chapters.

Exercises

2.1   **Alias Frequencies and the Principal Alias.** Standard alternating current is 60 Hz in the USA (50 Hz in many other countries) and radiates electromagnetic waves at this frequency, which are picked up by the wires used in experiments which act like antennas. It is difficult to exclude this radiated 60 Hz noise because every electrical wire that is not shielded (usually with a co-axial shield) receives or transmits the 60 Hz noise. Even co-axial cables are not always sufficient to suppress 60 Hz noise. However, if it is present, it can be identified by its characteristic sinusoidal variation at 60 Hz, assuming that it is properly sampled.

A. Find the largest digital sample interval that can properly sample 60 Hz, and also the associated sampling frequency and Nyquist frequency.

B. All frequencies above the Nyquist frequency will be aliased into a frequency that lies within the Nyquist band. If f is a frequency in the Nyquist band (the principal alias) then $f + nf_s$, where n is any positive or negative integer, will be an alias. For real-valued signals, positive- and negative-frequency values can be considered as the same, so if there is an alias at a negative frequency, it will also appear at the corresponding positive frequency. Find all the alias frequencies of 60 Hz in the interval $[-60, 60]$ Hz when the digital sample interval is 40 milliseconds.

2.2   **Statistics.** A computer random number generator is available on almost all computer systems for producing uniformly distributed independent random numbers on the interval $[0, 1]$. Using this, generate a matrix of 12,000 such numbers, after subtracting 0.5 from each, arranged in 12 rows by 1000 columns. These are now 12,000 numbers approximately uniformly distributed on the interval $[-0.5, 0.5]$, with zero mean. Consider each row of the matrix to be a time series. Write code functions that can calculate the mean, variance, and standard deviation of a time series. Refer to Appendix C for background reading, then do the following.

A. Show that the expected value of the variance of the time series on each row is 1/12, by finding the integral in the formal definition of an expectation (see Appendix C).

B. Find the variance of each of the 12 time series to confirm that the variance values are near 1/12, but not exactly because the values are computed from a finite number of data (1000~each). Find the average of these and show that it is even closer to 1/12, confirming the value of averaging to improve estimates.

C. Take the sum of 12 values down each column to form a new time series of length 1000, each value the sum of 12 values in respective columns. Find the variance of this series and confirm that it is close to 1. This confirms the property that, for independent random variables, the variance of their sum, which you have just determined, is the sum of their variances or (12 times 1/12). By the central limit theorem (Appendix C), the sum of the 12 time series is approximately Gaussian (normally) distributed. So, by summing 12 uniformly distributed time series (each uniformly distributed zero-mean white noise), you have created an approximately Gaussian-distributed zero-mean unit-variance white noise time series.

D. Use a histogram to create eight bins with boundaries $[-4, -3, -2, -1, 0, 1, 2, 3, 4]$ to show that about 1/3 (about 333 out of 1000) of the values in the series formed from the sum of 12 in part C fall into the $[-1, 0]$ or $[0, 1]$ bin, and that about 95 percent are located in the range $[-2, 2]$. The histogram suggests a bell-shaped curve, and the bin count shows the properties of Gaussian (normal) random variables with zero mean and unit variance, usually designated N[0, 1].

E. Convert your time series of Gaussian white noise so that it has mean 2 and variance 5. Verify these values with your mean and variance functions.

F. Create two series that are each the sum of 12 series of 1000 uniformly distributed zero-mean random numbers, so you have two approximately normally distributed unit-variance time series. Then square the value of each member of the two series and add the two series together. These now consist of 1000 approximately chi-squared-distributed random numbers, each with two degrees of freedom. (As explained in Appendix C, the number of degrees of freedom refers to the number of independent values added together. For example, a chi-squared random variable with two degrees of freedom is the sum of two independent squared Gaussian random variables, each with zero mean and unit variance.) Repeat with sums of respectively four and five squared series to obtain chi-squared series of 1000 chi-squared numbers with respectively four and five degrees of freedom. Then divide each series by its respective number of degrees of freedom (2, 4, 5) and plot them, offset vertically. This should show that chi-squared random variables with more degrees of freedom have less scatter. This result is used in Chapter 10 to justify averaging power spectrum estimates to reduce their scatter.

2.3   **Autocorrelation.** This should be done with paper and pencil to clarify how the autocorrelation is calculated and to verify how it differs for time series that are rough and smooth. For the time series $[-3, -2, -1, 0, 1, 2, 3]$ confirm by hand calculations that the autocorrelation is the same if the series is reversed in time.

2.4   **Decibels.**

A. When comparing the size of signals, electrical engineers generally accept that two signals are close to being the same if their magnitudes are within 1 decibel of each other. Thus engineers often say, "What is 1 dB among friends?" On the other hand, students unfamiliar with the decibel scale are more likely to say, "What is 1 dB?" Calculate a 1 dB increase in terms of a percentage change in amplitude.

B. One of the great controversies in marine mammal protection concerns the sound levels that whales and other marine mammals might be exposed to when seismic surveys are conducted in the ocean. Sound levels are always given in dB, but it turns out that the reference pressure for sound in water is not the same as that used for sound in air. Lawyers arguing these cases were completely ignorant of this fact, which caused great confusion for a period of years, leading to unjustified bans on scientific studies. Use the internet or other resource to find the two reference pressures used in air and in water.

3 Sinusoids and Fourier Series

In order to fully appreciate the Discrete Fourier Transform (DFT), a mainstay of geophysical data processing, which will be developed in the next chapter, we review here the Fourier series of continuous functions which represents analog signals as sums of sinusoids. In addition, the chapter provides a concise development of complex numbers, complex sinusoids, and complex number multiplication, all of which are essential to understanding the basic elements of the DFT, such as the distinction between positive and negative frequencies.

3.1 Sinusoids

Sinusoidal functions are oscillatory, repeating after an interval of time called the period, T. The reciprocal of the period is frequency, also called temporal frequency, denoted by the letter f. For example, the musical note "A" on a piano produces a sinusoidal variation of pressure at the frequency $f = 440$ Hz, so the period T is about 2.3 milliseconds. A related quantity, $2\pi f$, is the angular frequency, denoted by the Greek letter ω. If the period is measured in seconds then the temporal frequency is in cycles per second, or Hz. The special case $f = 0$, zero frequency, is often called DC, standing for direct current, a reference to electrical engineering applications. Signals that are DC are constant in time, with infinite period. The zero-frequency (DC) component of a time series corresponds to its mean or average value. With the exception of zero frequency, each sinusoidal constituent that contributes to an analog signal includes both cosines and sines, $c\cos(2\pi ft) + s\sin(2\pi ft)$, where c and s are respectively the cosine and sine component amplitudes. Cosines and sines are essentially the same function, shifted in time by one-quarter of the period $T = 1/f$. A sum of a cosine and sine of the same frequency can be expressed either as a shifted cosine or a shifted sine of that frequency with amplitude equal to the square root of $c^2 + s^2$.

Spatial variations can be described equally well by sinusoidal functions. For example, if the position along a profile is measured in meters, the distance between the sinusoidal peaks is called the wavelength. The spatial frequency, the reciprocal of the wavelength, then has units of cycles per meter and is called the wavenumber, although in some contexts the spatial frequency multiplied by 2π may be referred to as the wavenumber.

The amplitude of a sinusoid is half the peak to trough difference. The phase of a sinusoid refers to the location of its peak relative to a reference time, usually taken to be zero time. Phase may be expressed as a fraction of one period, which corresponds to either 2π radians or 360 degrees. For example, as shown in Figure 3.1, if the peak of a sinusoid occurs at time $t = T/8$, the phase at this point in time could be given as $T/8$, or as $\pi/4$ or 45 degrees. A phase of π radians (or 180 degrees) implies

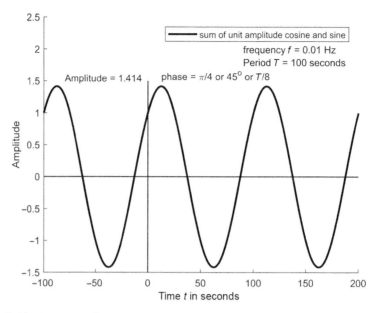

Figure 3.1 Sinusoids are described by their period or frequency, by their amplitude, and by their phase. In this example equal-amplitude cosine and sine functions are added together to make a sinusoid whose amplitude is the square root of the sum of squares of the amplitude of each, and whose phase is one-eighth of a period. That is, the phase refers to the location of the peak after time zero. The phase is altered by changing the relative amounts of the sine and cosine contributions. More cosine would reduce the phase (moving the peak to the left), for example.

that the first peak of the sinusoid is exactly one half-period past time zero, so a trough rather than a peak appears at time zero. Thus the common expression "changing the phase by 180 degrees" means reversing the sign of a sinusoid or of an entire function (by reversing the sign of all the sinusoids used to represent it).

3.2 Fourier Series

A continuous function of time $x(t)$ (an analog signal) defined over a finite time interval of length N can be represented as a sum of sines and cosines of Fourier frequencies, the set of all frequencies which have an integer number of cycles within the interval. We call the interval length N the record length. Fourier frequencies include zero frequency, one cycle per record length, two cycles per record length, and so on, up to an infinite number of cycles, that is, infinite frequency. Figure 3.2 shows the first few of the infinite number for a record length equal to 100 seconds. The frequency $f = 0$ corresponds to the sinusoid $\cos(2\pi 0t) + \sin(2\pi 0t)$, giving it a constant value of 1 over the interval. All other frequencies have zero-mean value over the interval. The Fourier frequencies are specified via an integer m as the set $[f = m/N], m = [0, 1, 2, \ldots, \infty]$. The interval over which $x(t)$ is defined can be chosen to be either $[0, N]$ or $[-N/2, N/2]$. The choice is a matter of convention but the connection with the Discrete Fourier Transform (described in the next chapter) is best made using $[0, N]$. The other choice works best when discussing the relationship between Fourier series and the Fourier transform of continuous functions; this is done in Appendix B.

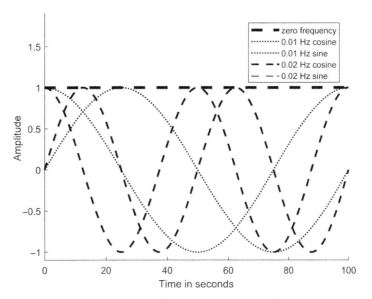

The first three Fourier frequencies for a time series of record length 100 seconds are zero frequency (constant value of 1) and one and two cycles per record length, with frequencies 0.01 and 0.02 Hz, respectively. Each Fourier frequency must allow for both a cosine and a sine function. Fourier frequency sinusoids are orthogonal to one another, so the integral of their product over the interval is zero.

The Fourier series of the analog signal $x(t)$ is

$$x(t) = c_0 + \sum_{m=1}^{\infty} \left[c_m \cos\left(2\pi \frac{m}{N}t\right) + s_m \sin\left(2\pi \frac{m}{N}t\right) \right] \tag{3.1}$$

and represents $x(t)$ within the interval $[0, N]$, except possibly at points of discontinuity, as discussed below. The Fourier series can be evaluated for times t outside the interval where it periodically replicates $x(t)$ at period N over the infinite time axis, as shown in Figure 3.3.

For a continuous function $x(t)$ the Fourier series includes an infinite number of both cosine and sine terms, represented by coefficients $[c_0, c_1, \ldots]$ and $[s_1, s_2, \ldots]$ for frequencies corresponding to integers $m = [0, 1, 2, \ldots]$. An infinite number of frequencies is required to allow for the situation where $x(t)$ may contain arbitrarily rapid variations, requiring very high values of frequency.

The Fourier series coefficients for frequency $f = m/N$ are found via integration over the interval $[0, N]$ of the function $x(t)$ multiplied by the sine or cosine at that frequency (see below). For the case $m = 0$ (zero frequency) there is only a cosine coefficient:

$$c_0 = \frac{1}{N} \int_0^N x(t)dt$$

The coefficient c_0 is the DC or mean value over the interval. The coefficient s_0 is zero, since $\sin(0) = 0$. For the other frequencies $m = [1, 2, \ldots, \infty]$ there are both sine and cosine coefficients;

$$c_m = \frac{2}{N} \int_0^N x(t) \cos(2\pi(m/N)t) \, dt$$

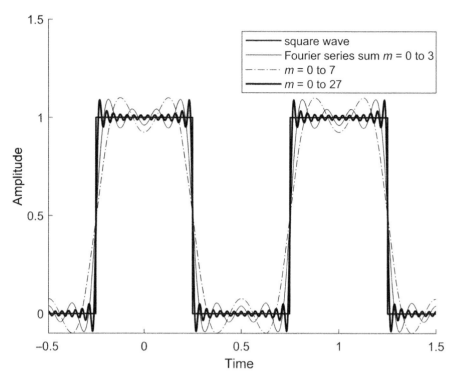

Figure 3.3 Partial Fourier series sums representing a function with a discontinuity reveal Gibbs oscillations. These become more concentrated near the discontinuity, and oscillate at a higher frequency as the number of terms in the series is increased. The analog function in this example is defined on the interval [0, 1], but the figure shows its periodic extension over a larger range, along with its partial Fourier series, which can be evaluated for any time inside or outside that interval.

and

$$s_m = \frac{2}{N} \int_0^N x(t) \sin(2\pi(m/N)t)\, dt$$

These expressions can be easily derived by assuming that a Fourier series exists for $x(t)$, and then using the orthogonality of cosines and sines at Fourier frequencies. To obtain the coefficients, first multiply the left- and right-hand sides of equation (3.1) by a particular sinusoid, say $\cos(2\pi(k/N)t)$ for some integer k. On the left-hand side we then have

$$x(t) \cos(2\pi(k/N)t)$$

and on the right side

$$c_0 + \sum_{m=1}^{\infty} [c_m \cos(2\pi(m/N)t) + s_m \sin(2\pi(m/N)t)] \cos(2\pi(k/N)t)$$

Now we integrate each of these expressions over the interval $[0, N]$. On the right-hand side there is an infinite number of integrals but all but one are zero, because the sines and cosines of different Fourier frequencies are orthogonal with respect to integration (meaning the integral of their product is zero) over the interval of length N. That is, every integral in the sum

$$\int_0^N \cos(2\pi(k/N)t) \cos(2\pi(m/N)t)\, dt$$

is zero when k and m are not equal, or when one function is a sine and the other a cosine. For the single non-zero integral when $k = m$ the above integral is equal to $N/2$, leaving the final expression

$$c_k = \frac{2}{N} \int_0^N x(t) \cos(2\pi(k/N)t)\, dt$$

3.3 Partial Fourier Sums

Students usually encounter Fourier series within the first two years of university-level calculus and related mathematics, where it is common to pay attention to convergence of the series to $x(t)$. Convergence in the interior of the interval is not particularly of interest because when $x(t)$ is an analog signal corresponding to a physical quantity it will be continuous. However, convergence at the endpoints of the interval is of considerable interest and is relevant to understanding both Fourier series and the Discrete Fourier Transform, which is a Fourier series for discretely sampled time series. Related to the question of convergence is the behavior of partial sums of Fourier series, when the series is truncated at a finite frequency. A signal $x(t)$ whose values differ at $t = 0$ and $t = N$ has a discontinuity when it is periodically extended onto an infinite time interval. A Fourier series contains continuous functions (sinusoids) and so is not able to represent this discontinuity. When $x(t = 0)$ and $x(t = N)$ are different, the Fourier series converges to the average value $[x(0) + x(N)]/2$. A partial sum of the series will show oscillations near a discontinuity. These Gibbs oscillations are more rapid and are concentrated more closely to the discontinuity as the number of frequencies is increased in a partial sum, as shown in Figure 3.3.

3.4 Complex Numbers

This section summarizes the properties of complex numbers as a prelude to expressing the Fourier series using complex sinusoids in place of real-valued cosines and sines. Despite their slightly menacing name, complex numbers and sinusoids greatly simplify calculations and notation in data processing. Students often encounter them in a mathematics course in high school or college, but there is rarely a clear application or purpose for them. In fact, they are essential elements of time series analysis and data processing, both aiding a conceptual understanding of many topics and also enabling rapid calculations via numerical algorithms such as the Fast Fourier Transform.

A discussion of complex numbers starts with the definition of the imaginary unit, the square root of -1, denoted by either i or j. Here we use i, but j is more common in some fields, such as electrical engineering. Some computer languages require a separate treatment of complex numbers.

Complex numbers arise naturally when one is trying to solve simple equations such as $u^2 + u + 1 = 0$. The quadratic formula gives two solutions (roots), $u = (-1/2) \pm i\sqrt{3}/2$. The two roots are complex, with the same real part $(-1/2)$ but imaginary parts of opposite signs. Numbers with the same real part but opposite signs for the imaginary part are complex conjugates of one another. The complex conjugate of a variable u is denoted by a superscript asterisk as u^*. In an expression for any complex number (even if i appears in several places), the complex conjugate is found by changing the sign of i everywhere in that expression. A number with zero for its imaginary part is pure real, and a number with zero for the real part is pure imaginary.

Real numbers can be associated with points on a number line, but complex numbers have separate real and imaginary parts, so there is a natural association with points in a plane. A complex plane is used to plot each complex number as a point, with the real part measured along the horizontal axis and the imaginary part along the vertical axis. This two-dimensional plot, called an Argand diagram, is a Cartesian coordinate system for displaying complex numbers. For every complex number $a + ib$ its Cartesian coordinates are the two real numbers a, b, and we call $a + ib$ the Cartesian form of a complex number. This form is probably most familiar but often not the most useful.

The correspondence between complex numbers and points in a plane suggests that a complex number can be described by a vector. Considering a vector to be an arrow lying within a plane, the tail of the arrow is at the origin and the head at the point $a + ib$. When adding complex numbers together, the rules for addition of vectors apply, placing the tail of one vector at the head of the other, and forming a vector sum. Equivalently, one adds separately the real and imaginary parts. Alternatively, a vector is described by both a magnitude and direction, and this leads to an equivalent representation of a complex number in terms of a magnitude or length of the vector multiplying a unit vector pointing in the direction of the vector. The polar form of a complex number gives these two quantities, the magnitude $\|a + ib\|$ (called the modulus) is the square root of the sum of the squares of the real and imaginary parts, $\sqrt{(a^2 + b^2)}$, the length of the arrow. The direction is the angle ϕ measured counterclockwise from the real axis. Thus $\tan\phi = b/a$, and so ϕ is the arctangent of the ratio of the imaginary and the real parts, $\phi = \tan^{-1}(b/a)$, as shown in Figure 3.4. To calculate ϕ we always use the computer function ATAN2(b, a) to find the arctangent, so that the angle will fall in the range $[-\pi, +\pi]$.

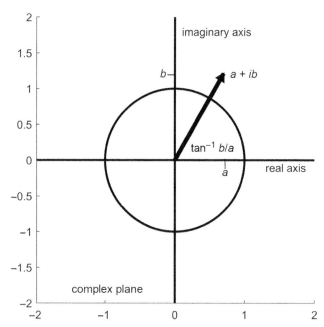

Figure 3.4 The complex number $a + ib$ is plotted as a point in the complex plane where the horizontal axis is used for the real part (a in this case), and the vertical axis for the imaginary part, b. A complex number can be thought of as a vector, with the addition of complex numbers corresponding one-to-one with vector addition. The unit circle is shown for reference.

To convert from the Cartesian to the polar form of a complex number, we first rearrange the Cartesian form:

$$a + ib = \|a + ib\| \times \left(\frac{a}{\|a + ib\|} + i\frac{b}{\|a + ib\|} \right) \tag{3.2}$$

Since $\cos(\phi) = a/\|a + ib\|$ and $\sin(\phi) = b/\|a + ib\|$, it follows that

$$a + ib = \|a + ib\| \times [\cos(\phi) + i\sin(\phi)]$$

which is the product of its length and a unit-length complex number (a unit vector) in the same direction. This Cartesian form is very nearly the polar form. Only one more step is needed, replacing the sine and cosine in $\cos(\phi) + i\sin(\phi)$ with a complex exponential. This step requires the Euler relationship, relating exponential and sinusoidal functions, which is

$$\exp(i\phi) = \cos(\phi) + i\sin(\phi) \tag{3.3}$$

For visual clarity, we will write the exponential function as $\exp(i\phi)$ rather than $e^{i\phi}$ throughout.

The Euler relationship is a simple but profound result that is easily verified using Taylor series expansions of sine ($\phi - \phi^3/3! + \phi^5/5! + \cdots$), cosine ($1 - \phi^2/2! + \phi^4/4! + \cdots$), and exponential ($1 + \phi + \phi^2/2! + \phi^3/3! + \cdots$) functions. The important conceptual result is that $\exp(i\phi)$ is a unit vector in the complex plane pointing in the direction given by the angle ϕ. Its modulus is unity and its direction is given by the angle ϕ with respect to the real axis.

Combining the previous two equations, we obtain the polar form of the complex number as

$$a + ib = \|a + ib\| \exp(i\phi)$$

All the usual properties of exponentials apply to complex numbers written in polar form. For example the natural logarithm of a complex number is

$$\ln(a + ib) = \ln[\sqrt{a^2 + b^2} \exp(i\phi))] = \ln[\sqrt{a^2 + b^2}] + i\phi$$

so the logarithm of a complex number is another complex number whose real part is the logarithm of the modulus, and whose imaginary part is equal to the angle $\phi = \tan^{-1}(b/a)$.

One of the most important reasons for using the polar form is to understand the multiplication of two complex numbers. Because exponents add upon multiplication, the product of two complex numbers has a modulus equal to the product of the two moduli, and an exponent (angle) equal to the sum of angles of the complex numbers being multiplied. This beautiful image of complex number multiplication as a scaling and rotation in the complex plane is obscure when the Cartesian form is used. We will find in Chapter 5 that complex numbers are perfectly suited to representing linear filter transfer functions because the scaling in magnitude describes the filter amplification and the rotation describes the filter phase response.

3.5 Complex Sinusoids

A complex exponential $\exp(i\phi)$ is a unit-modulus complex number, described as a vector pointing in the direction defined by angle ϕ. Therefore, provided that the angle ϕ is real, $\exp(i\phi)$ is a point on the unit circle in the complex plane. Now suppose that ϕ is not a constant but varies at a steady rate with time t, so that $\phi = 2\pi f t$, where f is frequency. The corresponding point on the unit circle

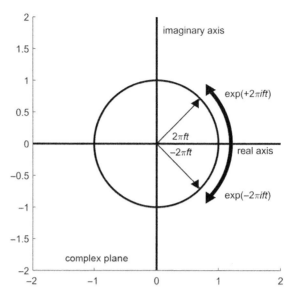

Figure 3.5 Complex sinusoids can be viewed as rotating vectors in the complex plane, counterclockwise for positive frequencies and clockwise for negative. The angle $2\pi ft$ changes linearly with time, making one revolution in the time period $T = 1/f$.

then changes in time, advancing around the circle at a steady rate in a counterclockwise fashion for positive f, making one rotation in the period $T = 1/f$. We have

$$\exp(i2\pi ft) = \cos(2\pi ft) + i\sin(2\pi ft)$$

This defines a complex sinusoid as a rotating vector in the complex plane, revolving at period $T = 1/f$, with the head of the vector tracing out points on the unit circle. The cosine part is the projection onto the horizontal (real) axis of the head of the vector, and the sine part is the projection onto the vertical (imaginary) axis.

If the sign of f is reversed then $\exp(i2\pi(-f)t) = \cos(2\pi ft) - i\sin(2\pi ft)$, because $\sin(-\phi) = -\sin(\phi)$ and $\cos(-\phi) = \cos(\phi)$. This clarifies the difference between positive- and negative-frequency f values. That is, positive-frequency sinusoids rotate counterclockwise in the complex plane, and negative-frequency sinusoids rotate clockwise. Figure 3.5 illustrates the sense of rotation of positive- and negative-frequency unit-amplitude sinusoids. The distinction between positive and negative frequencies allows us to define a frequency axis that is a real-number line extending from negative to positive infinity. The sampling theorem discussion noted that all alias frequencies are separated by positive- and negative-integer multiples of the sampling frequency $f_s = 1/\Delta t$. The distinction between positive and negative frequencies is now clarified by the use of complex sinusoids. On the frequency axis, the Nyquist band is the range of all frequencies between the positive and negative Nyquist frequencies.

We can add positive- and negative-frequency complex sinusoids to obtain the usual expression for the cosine:

$$\cos(2\pi ft) = \frac{\exp((2\pi i(+f)t)) + \exp((2\pi i(-f)t))}{2}$$

Thus a real-valued cosine function, used to describe real-valued analog signals, is the sum of two complex sinusoids of equal amplitude, one at frequency $(+f)$ and the other at $(-f)$. A similar result applies to the sine function:

$$\sin(2\pi f t) = \frac{\exp((2\pi i(+f)t)) - \exp((2\pi i(-f)t))}{2i}$$

Most common data processing problems deal with real-valued analog signals, so the corresponding Fourier series are usually given in terms of real-valued cosines and sines. However, the above identities show that cosines and sines are sums of complex sinusoids, with equal-magnitude contributions from positive and negative frequencies. The combination of positive and negative frequencies to obtain real-valued quantities is important in understanding the complex form of the Fourier series, and the Discrete Fourier Transform which follows.

A complex analog signal that contains entirely positive or entirely negative frequencies is an analytic signal. A three-dimensional plot of an analytic signal appears as a helical path rotating about a time axis, with the real and imaginary axes comprising the other two dimensions. Any purely real signal can become the real part of an analytic signal, with the imaginary part computed from the real part via a Hilbert transform. The theory underlying the Hilbert transform appears in Appendix B, and a computational algorithm for it is described in the next chapter. An analytic signal can be written in polar form as $A(t)\exp(i2\pi f(t)t)$ where $A(t)$ is the instantaneous amplitude and $f(t)$ is the instantaneous frequency. Since $A(t)$ varies in time, as does $f(t)$, the helical path changes amplitude and rate of rotation as it spirals around the time axis. The analytic signal concept is useful in understanding how information is encoded in radio broadcasts, with amplitude modulated (AM) broadcasts encoding information in $A(t)$ and frequency modulated (FM) broadcasts encoding information in $f(t)$. The analytic signal and related instantaneous amplitude and frequency time series are also used as signal attributes in applications other than radio. For example, the instantaneous frequency is commonly used as a signal attribute in the analysis of seismograms in exploration for oil and gas. A signal attribute is a new time series derived by a specified calculation from another time series. An application of the instantaneous-frequency attribute to the measurement of surface wave dispersion is given in the next chapter.

3.6 Chapter Summary

In preparing for a discussion of the Discrete Fourier Transform (DFT), it is important to have a clear understanding of Fourier series and of complex sinusoids and their properties. To illuminate these concepts the main points of the chapter have been as follows.

- A Fourier series represents an analog signal defined on an interval as a sum of sinusoids at Fourier frequencies; Fourier frequencies are those that have an integer number of cycles in the record length.
- Because the DFT is implemented entirely with complex numbers and complex sinusoids, a firm understanding of their properties is essential. Complex numbers may be thought of as vectors in the complex plane, and in polar form (which is the most useful representation) the modulus is the vector length and the exponent provides the direction (the angle with respect to the real axis).
- Complex number multiplication can be described as a rotation of a complex number (vector) in the complex plane and a scaling in magnitude.
- Positive- and negative-frequency sinusoids are distinct, with positive frequency corresponding to a counterclockwise rotation of a vector (representing a complex number) and negative frequency to clockwise rotation.

- A complex function consisting of entirely positive (or entirely negative) frequencies is complex-valued and is called an analytic signal. A purely real function can become the real part of an analytic signal via the Hilbert transform (described in Appendix B), which can be implemented numerically using the DFT as described in the next chapter.
- Although most geophysical signals are real-valued, complex sinusoids are used to represent them in the DFT and complex sinusoids are also used in the analysis of linear digital filters. To make a real-valued signal from complex sinusoids requires positive and negative frequencies of equal magnitude but opposite phase. In the case of complex Fourier series, this is termed Hermitian symmetry.

Exercises

3.1   **Sinusoids.** The sum $(\cos(2\pi ft) + s\sin(2\pi ft)$ is equal to either a pure cosine or pure sine of the same frequency with amplitude $A = [c^2 + s^2]^{1/2}$. An example is illustrated in Figure 3.1, where $c = s = 1$. Use complex-exponential expressions for both cosine and sine to find the phases p and q in the equations below. Do this by inserting complex-exponential expressions for cosine and sine, and selecting all positive-frequency terms on both sides of the equation, equating them to obtain expressions for A, p, and q:

$$c\cos(2\pi ft) + s\sin(2\pi ft) = A\cos(2\pi ft - p) = A\sin(2\pi ft - q)$$

3.2   **Complex Frequencies.** This exercise illustrates the computational utility of using complex numbers to simulate damping in oscillatory systems.

     The frequency of a sinusoidal oscillation is usually understood to be a real-valued quantity, given in Hz. If frequency is allowed to be complex-valued, it can be used to describe, as well as the actual frequency, the attenuation or dissipation in vibrating systems. Recognizing this, complex-valued frequencies are a useful computational tool to describe the dissipation of waves in numerical wave modeling codes, used in global and exploration seismology. Damping can also be introduced using complex-valued elastic parameters or complex wave speeds in these codes.

     To illustrate this, compute the output of a geophone after an abrupt disturbance. (A geophone is a high-frequency seismometer consisting of a mass on a spring with damping, which puts out a voltage using a moving coil in a magnetic field.) We can describe the geophone using the complex frequency $f = f_0(1 + i/2Q)$, where f_0 is the resonant frequency of the undamped mass and spring. A typical value for geophones is $f_0 = 12\,\mathrm{Hz}$. The quantity Q is known as the quality factor, a dimensionless positive number larger than 1 which can be adjusted in most geophones by introducing a resistor across the output coil. Without damping, the output in terms of a complex sinusoid is $\exp(i2\pi f_0 t)$. Let t be an array of times at 0.01 second intervals for 1 second. Plot the real part of $\exp(i2\pi f_0 t)$ versus time for a complex frequency when $Q = 2, 5, 10, 25$ for $f_0 = 12\,\mathrm{Hz}$. Your graphs should show damped sinusoidal oscillations and that lower Q means more damping.

3.3   **Complex Numbers.** Here are a few simple activities to illustrate computations with complex numbers and to demonstrate numerical precision limits.

A. There are generally N different Nth roots of the number 1. Use the polar form $1 = \exp(i2\pi)$ to show that all N roots are spaced at equal angles around the unit circle in the complex plane. Find the three cube roots of the number 1.

B. For the two complex numbers in Cartesian form $2 + 3i$ and $1 + 4i$, compute their product in two ways – first using multiplication of the Cartesian forms, and second after first expressing them in polar form, taking the square root of the sum of real and imaginary parts to find the modulus of each complex number, and using the arctangent function (ATAN2) to find the phase angle in radians. After multiplying magnitudes and adding phases, convert back to Cartesian form. Demonstrate in these computations that the magnitude of the difference between the two ways of doing the multiplication is not exactly zero, reflecting the limited precision of floating point arithmetic, as discussed in Chapter 2.

C. Find the complex number R that, when used as a multiplier, produces a clockwise rotation of an angle of the vector associated with complex number C; C is a vector whose tail is at the origin, and head at the point $(\text{real}(C), \text{imag}(C))$ in the complex plane.

D. The Euler relationship is easily verified using Taylor series expansions of sine ($\phi - \phi^3/3! + \phi^5/5! + \cdots$), cosine ($1 - \phi^2/2! + \phi^4/4! + \cdots$), and exponential ($1 + \phi + \phi^2/2! + \phi^3/3! + \cdots$) functions. Show that these agree with the Euler relationship by combining the first three terms of the sine and cosine series to show that these equal the first six terms of the exponential when the function is evaluated for $i\phi$.

3.4 Orthogonality of Fourier Frequency Sinusoids. The orthogonality of functions means that the integral of their product over an interval is zero. The orthogonality of time series or vectors means that their dot product (the sum of term-by-term products) is zero.

A. A continuous function is defined over a time interval of length 100 seconds. When deriving a Fourier series to represent it, the orthogonality of sines and cosines of Fourier frequencies is invoked, meaning that the integral of the product of two functions over the interval is zero. Show that cosines of frequencies 0.01 Hz and 0.02 Hz are orthogonal by computing this integral.

B. Discretely sampled Fourier frequency sinusoids are also orthogonal. In this case the dot product of the arrays of sampled sinusoids is computed. Orthogonality then means that the dot product (the sum of term-by-term products) is zero. Show by a computational example that two vectors of 100 samples, for integer $t = [0, 99]$ seconds, of $\cos(2\pi f_1 t)$ and $\cos(2\pi f_2 t)$ are orthogonal, when f_1 and f_2 are two different Fourier frequencies that are integer multiples of 0.01 Hz, the fundamental Fourier frequency. Also verify orthogonality when one is a cosine and the other a sine.

C. A square function $b(t)$ defined on the interval $[-0.5, 0.5]$ equals 1 in the range $[-0.25, 0.25]$ and zero outside this subinterval. This is called a boxcar because it resembles a railroad boxcar in silhouette. Periodically extended, it would appear on the interval $[0, 1]$ with a value of 1 up to $t = 0.25$, zero in the interval $[0.25, 0.75]$, and 1 for $t = [0.75, 1]$. It is an even function so has a Fourier series representation with no sine coefficients (sine is an odd function). The coefficients are found to be, upon integration

$$c_0 = \frac{1}{2}$$

and

$$c_m = \frac{2}{m\pi}\sin(m\pi/2)$$

Using densely spaced values of t, plot partial Fourier sums for values $m = [1, 3, 5, 7, 9]$ to show the general properties of Gibbs oscillations at points of discontinuity. This is an approximate replication of Figure 3.3.

4   The Discrete Fourier Transform

The Discrete Fourier Transform (DFT) is one of the most important tools in geophysical data processing and in many other fields. The DFT may be understood from a number of viewpoints, but here we emphasize that it is a Fourier series representing a uniformly sampled time series as a sum of sampled complex sinusoids. We refer to the time series as being in the time domain while the set of its complex-valued sinusoidal coefficients computed using the DFT is in the frequency domain. The inverse DFT (IDFT) computes time series values by adding together Fourier frequency sinusoids, each scaled by a frequency domain sinusoidal coefficient. We develop the DFT by converting the ordinary Fourier series to complex form, transitioning to sampled time series, and finally explaining standard normalization and the usual (and often baffling) frequency and time ordering conventions. The DFT came into widespread use only in the 1960s after the development of Fast Fourier Transform (FFT) algorithms. The speed of FFT algorithms has led to many important applications. Those presented here include the interpolation and computation of analytic signals for real-valued time series. In later chapters we show the important roles of the DFT in linear filtering and spectral analysis.

4.1   Fourier Series in Complex Notation

We can convert the ordinary Fourier series, discussed in Section 3.2, to complex notation, replacing the real sine and cosine functions with complex sinusoids. This is an essential step in understanding both the Fourier transform in the continuous-time case and the Discrete Fourier Transform. It also generalizes the Fourier series concept to allow for the possibility that $x(t)$ is a complex-valued analog signal. From the previous discussion we anticipate that the moduli of the positive- and negative-frequency components will be the same if $x(t)$ is a real-valued signal.

Assume for the moment that $x(t)$ is real-valued, and define a complex Fourier series coefficient X_m in terms of the real coefficients c_m and c_s (see Section 3.2) as $X_m = [c_m - is_m]/2$ for positive values of m (positive frequencies $+ m/N$) and $X_{-m} = [c_m + is_m]/2$ for negative values of m, with $f = -m/N$. For the case $m = 0$, the coefficient is simply $X_0 = [c_0]$. Therefore, the positive- and negative-frequency coefficients, X_m and X_{-m}, are complex conjugates of each other when $x(t)$ is real-valued, and, for this case, when the positive- and negative-frequency coefficients are added together the imaginary parts will cancel and the result will be purely real. When the positive- and negative-frequency coefficients are conjugates of one another, the coefficients are said to have Hermitian symmetry.

If we write down the sum of X_m and X_{-m} multiplied by their corresponding exponential sinusoids (see Section 3.5), respectively setting $f = m/N$ and $f = -m/N$, we obtain

$$[(c_m - is_m)/2] \exp(2\pi i(m/N)t) + [(c_m + is_m)/2] \exp(-2\pi i(m/N)t)$$

On expanding out the exponentials and gathering up terms, we see that this sum is the term in the real-valued Fourier series in Section 3.2 for $f = m/N$, that is

$$c_m \cos(2\pi(m/N)t) + s_m \sin(2\pi(m/N)t)$$

We deduce that the complex form of the real-valued Fourier series in Section 3.2 is then, extending the summation from $-\infty$ to $+\infty$ to take account of the positive and negative frequencies,

$$x(t) = \sum_{m=-\infty}^{\infty} X_m \exp(2\pi i(m/N)t) \tag{4.1}$$

In the same way, we can convert the formulas at the end of Section 3.3 for finding c_m and s_m from $x(t)$ to obtain a remarkably simple result that is correct for all frequencies, including zero frequency:

$$X_m = \frac{1}{N} \int_0^N x(t) \exp(-2\pi i(m/N)t)\, dt \tag{4.2}$$

This result is also correct when $x(t)$ is a complex-valued analog signal, but in that case the positive- and negative-frequency coefficients do not have Hermitian symmetry.

4.2 From Fourier Series to DFT

The DFT is a truncated Fourier series expressed in complex notation. Analog signals that have been sampled by adhering to the sampling theorem (Section 2.2) contain no frequencies higher than the Nyquist. Therefore, their Fourier series representation can be truncated at the Nyquist frequency, because all higher frequency coefficients are zero. Upon conversion to a complex-valued Fourier series, only frequencies between negative and positive Nyquist are present.

To develop these ideas, we first define the sampling frequency $f_s = 1$, implying that $\Delta t = 1$, and the Nyquist band is then the range of frequencies $[-1/2, 1/2]$. The sampling time values are integers $t = [0, 1, \ldots, N-1]$, and the record length N is an integer. For the development N is taken to be an even integer, but the case where N is odd differs only slightly. Integer values of t may be converted to seconds and frequencies to Hz once the actual sample interval is specified. The computation of the DFT requires that t be an integer.

The IDFT (inverse DFT) uses Fourier series coefficients to obtain discrete time samples, so transforms the set of complex coefficients from the frequency domain to the time domain. Assume for the moment that the Fourier coefficients X_m are known. To obtain the values of the continuous signal $x(t)$ at the sample times, $x_t = [x_0, x_1, \ldots, x_{N-1}]$, we simply compute the Fourier series in complex form obtained in the previous section:

$$x_t = \sum_{m=-N/2+1}^{N/2} X_m \exp(2\pi i(m/N)t)$$

but evaluating the sum on the right-hand side only at the times of the N samples $t = [0, 1, \ldots, N-1]$.

The truncated Fourier series sum given above does not include the negative Nyquist frequency, but instead begins with $m = -N/2 + 1$, continuing to frequencies up to the positive Nyquist where $m = N/2$. Because t is an integer, the negative and positive Nyquist frequency sinusoids are identical, so only one needs to be included.

The forward transform converts time series values into a set of Fourier coefficients. While in the continuous case the Fourier series coefficients are found via integration of the products of $x(t)$ with sinusoidal functions (see the end of Section 3.2), now only discrete samples $x_t = [x_0, x_1, \ldots, x_{N-1}]$ are available. Therefore the integral used in the continuous case,

$$X_m = \frac{1}{N} \int_0^N x(t) \exp(-2\pi i(m/N)t)\, dt$$

must be replaced with a sum for the discrete values in the linear array x_t. Intuition suggests replacing dt with $\Delta t = 1$, so that, using summation in place of integration, we obtain

$$X_m = \frac{1}{N} \sum_{t=0}^{N-1} x_t \exp(-2\pi i(m/N)t)$$

This intuition is correct, so we now have a definition of the DFT that calculates complex Fourier series coefficients X_m from time series values x_t. A slight modification of the DFT definition, to conform with contemporary normalization conventions, removes the factor $1/N$, as will be discussed in Section 4.4.

4.3 Frequency and Time Ordering

A final step is to replace the range of frequencies for the IDFT from $m = [-N/2 + 1, \ldots, N/2]$ to $m = [0, 1, \ldots, N-1]$. While using the values $m = [N/2+1, \ldots, N-1]$ appears to include frequencies above the Nyquist, we can show that this range actually includes all the negative frequencies arranged in reverse order. That is, frequencies corresponding to $m = [-N/2 + 1, \ldots, N/2]$ are found using $m = [0, 1, \ldots, N-1]$ but arranged as

$$f = [0, \text{positive } f, \text{Nyquist, negative } f \text{ in reverse order}]$$

To verify this, note that Fourier frequency sinusoids are unchanged if multiplied by unity. In particular, the factor $\exp(i2\pi(tN/N)$ is equal to unity because t is always an integer. When $m = -N/2 + 1$, the negative frequency sinusoid

$$\exp(i2\pi t(-N/2 + 1)/N)$$

when multiplied by $\exp(i2N\pi t/N)$ becomes

$$\exp(i2\pi t(N/2 + 1))$$

which is the frequency associated with $m = N/2 + 1$. By similar arguments this reordering applies to all the other negative frequencies.

A similar dual interpretation of time values is also possible. To understand this, recall that an odd continuous function $(x(-t) = -x(t))$ has a Fourier series containing only sine terms because sine is an odd function. If $x(t)$ is even $(x(-t) = x(t))$ its Fourier series must contain only cosines. In the case of the DFT, if all X_m values are purely real, then it must correspond to an even time series because the sine coefficients (in the imaginary part of X_m) are all zero. Although x_t was originally defined only at positive times, we can show by an example how even and odd time series will appear. For $N = 8$ an example of an even series is $x_t = [1, 1, 1, 0, 0, 0, 1, 1]$, because for the purpose of DFT calculation the associated times are $[0, 1, 2, 3, \pm 4, -3, -2, -1]$, so that the values at positive and negative times are equal. An example of an odd series is $[0, 1, 1, 0, 0, 0, -1, -1]$; again the associated times are $[0, 1, 2, 3, \pm 4, -3, -2, -1]$ but now at negative times the values are the negative of those at positive times. An odd time series will always have a zero value associated with time zero and a zero value associated with the end time, which is ±4 in this case.

For real x, the Hermitian symmetry of X is illustrated in the following example for record length 8:

$$x = [x_0, x_1, x_2, x_3, x_4, x_5, x_6, x_7]$$

$$DFT(x) = X = [X_0, X_1, X_2, X_3, X_4, X_5, X_6, X_7]$$

The Fourier frequencies are

$$f = \left[0, \frac{1}{8}, \frac{2}{8}, \frac{3}{8}, \frac{\pm 4}{8}, -\frac{3}{8}, -\frac{2}{8}, -\frac{1}{8}\right]$$

The Nyquist frequency for $N = 8$ is in array position $5 = N/2 + 1$, and can be interpreted as either positive or negative 1/2. Here X_0 (zero frequency) and X_4 (Nyquist) are pure real; X_1 is the conjugate of X_7; X_2 is the conjugate of X_6; X_3 is the conjugate of X_5; and X_4 is the conjugate of itself (therefore real-valued). The Hermitian symmetry for odd-length time series is similar except that the Nyquist frequency is absent.

4.4 DFT Normalization Conventions

The form of the forward and inverse transforms is identical except for the sign of the exponential and the placement of the normalization factor $1/N$, so various normalizations and definitions of DFT and IDFT can be adopted. When DFT algorithms were first developed, the same code was used for both forward and inverse transforms. The user specified whether a positive or negative exponent defined the forward transform, and where to place normalizing factor $1/N$. This factor can be assigned to the forward transform or to the inverse transform, or a factor $1/N^{1/2}$ can be applied to both forward and inverse transforms. Two widely used contemporary computing environments (MATLAB and python) apply $1/N$ to the inverse transform, and we adopt this convention. The final forms of the IDFT and DFT are then respectively

$$x_t = \frac{1}{N} \sum_{m=0}^{N-1} X_m \exp(2\pi i(m/N)t$$

and

$$X_m = \sum_{t=0}^{N-1} x_t \exp(-2\pi i(m/N)t)$$

4.5 Sinusoidal Coefficients of a Climate Time Series

With our adopted normalization, the DFT values and Fourier series coefficients are related by

$$X_m = N[c_m - is_m]/2$$

A simple but useful application of the DFT is to find the sinusoidal constituent at a frequency of interest. The real part of the complex-valued DFT yields the cosine coefficient (after multiplying by 2 and dividing by N), while the imaginary part provides the sine. However, this gives a correct result only if the frequency of interest is a Fourier frequency. For example, it is common to compute the average annual variation of a climate parameter, such as temperature, from many years of data. The frequency of interest is 1 cycle per year, which will be a Fourier frequency if the record length is precisely an integer number of years. If not, the DFT can still be used by truncating the series length to an integer number of years. If the frequency of interest is not a Fourier frequency, then sinusoidal coefficients must be found by least squares, as described in Chapter 7. Least squares provides a solution for any length record, and it works with missing data, data of variable quality, multiple samples at the same time, and irregularly sampled data. For a uniformly sampled time series whose length is an integer number of years, the annual sinusoidal coefficients found using either the DFT or least squares are the same.

Figure 4.1 shows how the seasonal component of a climate time series can easily be determined using the DFT. The sea level variations at Victoria, British Columbia, are monthly values for exactly 29 years, an array of length 348. The frequency spacing in the DFT is therefore the reciprocal of the

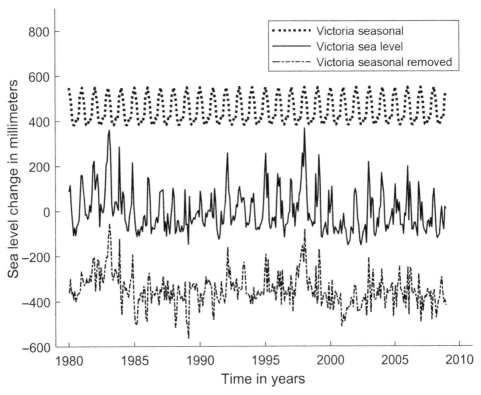

Figure 4.1 The DFT is useful for finding cosine and sine coefficients at Fourier frequencies. An important application is to find annual (1 cycle per year (cpy)) or seasonal (1, 2, 3, … cpy up to the Nyquist frequency) variations. The DFT can be used when the time series is exactly an integer number of years in length and there are no missing data. If this is not the case, then least squares is used, as described in Chapter 7. Here, a 29-year time series of monthly average sea level variations at Victoria, British Columbia (also given in Figure 1.3) is analyzed. Here the middle curve is the original time series with 12 samples per year. The top curve is the seasonal component obtained by setting all DFT values to zero except those at the seasonal frequencies, then computing the IDFT. Array locations for seasonal frequencies are given in the text. Both positive- and negative-frequency values must be included to preserve Hermitian symmetry. The bottom curve is the non-seasonal residual, obtained by setting to zero all seasonal frequency values in the DFT and then computing the IDFT. By removing such frequencies, the DFT can be used to perform linear filtering in the frequency domain.

record length or 1/29 cycles per year (cpy). Then the DFT values at array positions [30, 59, 88, 117, 146, 175, 204, 233, 262, 291, 320] correspond to positive frequencies [1, 2, 3, 4, 5, 6] cpy and the negative frequencies [−5, −4, −3, −2, −1] cpy. These are harmonics of 1 cpy, which contribute to a variation that repeats each year; 6 cpy is the Nyquist frequency. Removing the seasonal component requires only setting these values in the DFT array to zero, as is done in Chapter 10, to precondition this and a similar series prior to computing the spectrum of coherence (Figures 10.9 and 10.10).

To find the cosine and sine coefficients we scale the real and imaginary parts of the complex coefficients by the factor 2/348. The origin time for the cosine and sine coefficients is that of the first value of the time series, nominally mid-January. To express the cosine and sine coefficients for a reference time of January 1 would require that they be adjusted. The actual numerical values of the coefficients are given in the following table (the units in the second and third columns are millimeters) and show that the seasonal component is mostly due to 1 and 2 cpy terms:

cpy	cosine	sine
1	78.3471	−13.6157
2	22.9943	2.0008
3	−7.2011	−4.0287
4	−5.7586	3.4442
5	6.2679	1.3456
6	8.8391	0.0000

4.6 FFT Algorithms

While the DFT has been known for a long time, its importance as a data processing tool began with the publication of efficient algorithms known as FFTs (Fast Fourier Transforms) in the 1960s. The ideas underlying FFTs (some attributed to Gauss in the nineteenth century) preceded the development of digital computers. FFT algorithms are fast when the record length N is a highly composite integer, meaning that it has many factors. It has the most factors when it is an integral power of 2, so the first FFT algorithms were written for $N = 2, 4, 8, 16, 32, 64, \ldots$, but fast algorithms have been developed for other record lengths so there is no practical restriction on N.

While we derived the DFT using its connection with the Fourier series, two other interpretations are useful in understanding the ideas that have guided the development of FFT algorithms. One is that the DFT is the evaluation of a polynomial $X(Z)$, the Z transform of x_t, formed using time series values as coefficients of integer powers of Z in the following way:

$$X(Z) = x_0 + x_1 Z^1 + x_2 Z^2 + \cdots + x_{N-1} Z^{N-1}$$

The Z transform is useful in many other contexts and will reappear in later chapters. For the moment $X(Z)$ might be evaluated for any real or complex Z value. However, with our adopted convention, placing the normalization factor $1/N$ with the IDFT, the DFT evaluates the Z transform for Z values equally spaced around the unit circle (in the complex Z plane). That is, it evaluates the Z transform for values $Z = \exp(-i2\pi f)$ for Fourier frequencies $f = [0, 1/N, 2/N, \ldots]$. An FFT algorithm can then be understood as an efficient way to evaluate this polynomial at these equally spaced locations. Further discussion of the complex Z plane is found in Chapter 8.

The DFT can also be considered as a linear transformation between two vectors, transforming the column vector $x = x_t$ in the time domain to $X = X_m$ in the frequency domain. A transformation matrix V accomplishes this via $X = Vx$. An FFT algorithm may then be understood as a method of factoring V to reduce computational effort. The transformation between the two vector spaces is length-preserving, as expressed by the Parseval theorem, which states that the sum of the squared values (moduli) in the array x_t equals the sum of the squared moduli of X_m, with suitable normalization involving the record length N. The Parseval theorem is used later in applications of the DFT to spectral analysis.

As an example, when $N = 4$, the matrix V (a Vandermonde matrix) is

$$\begin{bmatrix} 1 & 1 & 1 & 1 \\ 1 & v & v^2 & v^3 \\ 1 & v^2 & v^4 & v^6 \\ 1 & v^3 & v^6 & v^9 \end{bmatrix}$$

where $v = \exp(-2\pi i/4)$. The inverse of V implements the IDFT, reversing the sign of the exponent. Thus, the inverse of V is its complex conjugate, with normalization factor $1/N = 1/4$:

$$\frac{1}{4}\begin{bmatrix} 1 & 1 & 1 & 1 \\ 1 & v^{-1} & v^{-2} & v^{-3} \\ 1 & v^{-2} & v^{-4} & v^{-6} \\ 1 & v^{-3} & v^{-6} & v^{-9} \end{bmatrix}$$

4.7 Zero-Padding and Interpolation

A useful DFT application is to trick the FFT algorithm into evaluating the Z transform at more densely spaced points on the unit circle in the Z plane by adding zeros to the end of the series. When k zeros are added (called zero-padding), the new series is called $pad_k(x_t)$. For example, if $k = N$, the series length is doubled but the added zeros make no contribution to the DFT summation. However, the Fourier frequencies of the padded series are $f = [0, 1/2N, 2/2N, 3/2N, \ldots]$ so that every other value is a Fourier frequency of the original x_t. Thus padding N zeros in the time domain has interpolated to midpoints in the frequency domain.

A corresponding conclusion is that padding zeros in the frequency domain will interpolate in the time domain. This yields a very useful method of interpolation. Not only is it efficient (because FFT algorithms are fast), but the interpolated time series is band-limited, containing no frequency above the original Nyquist. Of the many other interpolation methods (cubic spline, linear, and others), none has this band-limited property, which is important in certain applications.

To implement DFT interpolation we pad X_m with zeros and then compute the IDFT. For real x_t, zero-padding must preserve Hermitian symmetry, as is illustrated by a simple example. Suppose we want to do midpoint interpolation of a time series of length 8. The DFT is calculated first:

$$DFT(x_t) = [X_0, X_1, X_2, X_3, X_4, X_5, X_6, X_7]$$

Now we pad zeros above the Nyquist, retaining Hermitian symmetry by splitting the Nyquist value into two halves and placing seven zeros between the split Nyquist values to yield 16 terms. (For an odd-length series it is not necessary to split the Nyquist value, because it is not computed.) Then the frequency domain values corresponding to midpoint time domain interpolation are

$$X^{int} = [X_0, X_1, X_2, X_3, X_4/2, 0, 0, 0, 0, 0, 0, 0, X_4/2, X_5, X_6, X_7]$$

When $IDFT(X^{int})$ is computed it must be rescaled by a factor 2 because, with our adopted normalization, a factor $1/16$ is applied in computing the IDFT of a length-16 series, while $1/8$ is the correct factor, because the original time series was of length 8:

$$x_t^{int} = \frac{1}{8} \sum_{m=0}^{15} X_m^{int} \exp(2\pi i(m/16)) = 2 \times IDFT[X^{int}]$$

The interpolated values of x_t^{int} are at times $t = [0, 1/2, 1, 3/2, \ldots]$, midway between the original $t = [0, 1, 2, \ldots]$. The last value in x^{int} is an interpolation between the first and last values of x_t and is assigned to time $t = N - 1/2$.

4.8 DFT Interpolation Example

An example application of the DFT to interpolation appears in Figure 4.2. The amplitude of the original discrete samples increases linearly over time. There is a sharp discontinuity between the first and last samples, creating a discontinuity in a Fourier series (that is, a DFT) representation. Normally this would not be a good candidate for DFT interpolation because of this discontinuity, but it is chosen to show that DFT interpolation involves computing, at more finely spaced times, values of a Fourier series that has been truncated at the Nyquist frequency. As a truncated Fourier series, it will contain Gibbs oscillations, which are the ripples seen throughout the interpolated series in Figure 4.2.

4.9 Analytic Signal Computation and Application to Measuring Surface Wave Dispersion

As noted near the end of Section 3.5, every real-valued signal has a companion 90 degree phase-shifted version, its Hilbert transform. (A discussion appears in Appendix B.) For example, $\cos(2\pi ft)$, phase-shifted by 90 degrees has a Hilbert transform of $\pm \sin(2\pi ft)$. (The sign conventions for the Hilbert transform vary.) Therefore, the analytic signal $\exp(i2\pi ft)$ corresponding to $\cos(2\pi ft)$ has real part equal to cosine and imaginary part equal to a sine: $\cos(2\pi ft) + i \sin(2\pi ft) = \exp(i2\pi ft)$. This polar form of the analytic signal shows that the instantaneous amplitude is constant (equal to 1) and the instantaneous frequency f is also constant. An analytic signal can similarly be defined for any time series x_t, and requires a computational algorithm. The DFT provides a simple one, as follows.

Let X_m^a denote the DFT of the discrete analytic signal x_t^a corresponding to the real-valued time series x_t. First compute X_m, the DFT of x_t. Then set zero-frequency and Nyquist-frequency values in X_m^a equal to those of X_m. All positive-frequency values in X_m^a are set to be twice those of X_m, while all negative-frequency values are zero. The IDFT of X_m^a is the discrete analytic digital signal

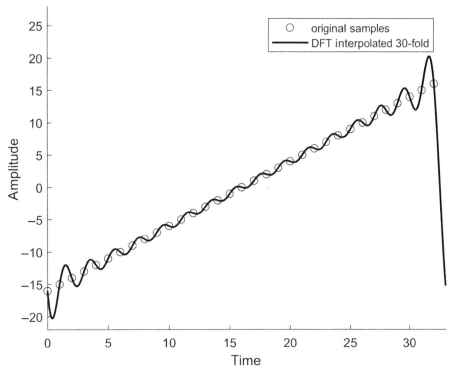

Figure 4.2 A discretely sampled ramp function whose samples are indicated by circles is to be interpolated using the DFT. When the time series is extended periodically, there is a large difference beween the first and last values. The densely interpolated result shows oscillations at the time series endpoints. These are the Gibbs oscillations associated with a partial Fourier series sum, as shown in Chapter 3. The DFT is just such a partial Fourier sum, truncated at the Nyquist frequency, and the ramp function, with a sharp discontinuity when periodically extended, is chosen to illustrate this behavior. DFT interpolation involves padding zeros in the frequency domain in order to evaluate the IDFT at more densely spaced times. Its advantages are that the interpolated values are band-limited to the original Nyquist frequency and that the method is computationally efficient because it uses an FFT algorithm.

x_t^a whose real part is x_t and whose imaginary part is its numerical Hilbert transform. Written in polar form the analytic signal (discrete time series) is

$$x_t^a = |x_t^a| \exp(i2\pi(f_t)t)$$

The two associated time series are $|x_t^a|$, the instantaneous amplitude, and the instantaneous frequency f_t, which is the time derivative of the phase $2\pi(f_t)t$ divided by 2π. An approximate derivative using the differences between adjacent samples is commonly used. Further discussion of such derivative filters appears in Chapter 8. One difficulty in computing the derivative of the phase is that numerical values (calculated using the ATAN2 function) are initially restricted to the range $[-\pi, \pi]$. This creates discontinuities in the instantaneous phase at the $\pm\pi$ boundaries. These discontinuities need to be removed before a time derivative can be calculated numerically, a process called phase unwrapping. Many computational environments include algorithms for phase unwrapping.

A geophysical application of the analytic signal is illustrated in Figures 4.3 and 4.4. Figure 4.3 is the horizontal transverse component of waves received at the seismic station PAYG in the

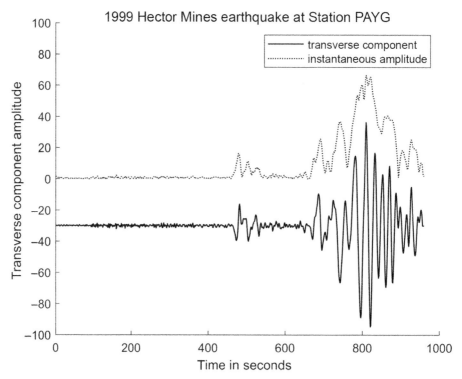

Figure 4.3 A portion of the Hector Mines transverse horizontal component observed at station PAYG from Figure 1.6. The dominant arrival consists of Love waves, which are shear horizontal (SH) waves trapped near the surface by the effects of increasing shear velocity with depth. The DFT was used to compute the analytic signal via a numerical Hilbert transform algorithm, as described in the text. The analytic signal modulus is called the instantaneous amplitude, and is plotted, offset, above the seismogram. It can be understood as an envelope function, giving a continuous non-negative amplitude variation with time.

Galapagos Islands due to the 1999 Hector Mines earthquake in southern California. The first-arriving P waves (compressional waves) make very little contribution to this component, but the large waves arriving after around 700 seconds are Love waves, transverse horizontal shear surface waves that are dispersed. Dispersion causes the longer-period waves to arrive before the shorter-period waves, reflecting the effects of increasing shear velocity with depth. The instantaneous amplitude gives the envelope of the wave amplitude, as shown in Figure 4.3. The instantaneous frequency allows us to quantify the arrival time of waves of different frequencies. An approximate dispersion calculation measures the time between two adjacent peaks, to determine the period or frequency of the wave. The arrival time of the wave can be taken to be midway between the two peaks. However, the instantaneous frequency method provides a better approach, giving a continuous variation of the frequency of arriving waves in place of the two or three values that one could obtain from measuring the peak to peak times. Figure 4.4 shows a portion of the Love waves, with the instantaneous frequency displayed below. There is a steadily increasing frequency of the arriving waves, meaning that higher-frequency waves travel more slowly. Lower-frequency (longer-period, longer-wavelength) surface waves travel faster because a larger proportion of their particle motion is at greater depths, where shear wave speeds are higher. The arrival times of the different-frequency Love waves are used to find group velocity curves, giving the apparent velocity of waves of different

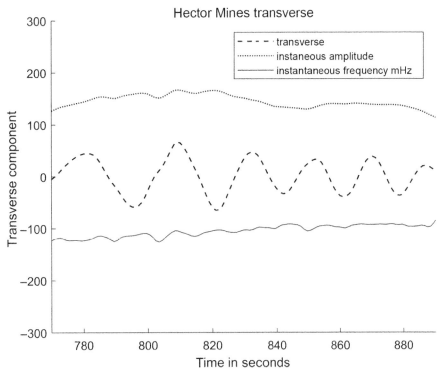

Figure 4.4 A portion of the Love wave arrivals from the previous figure shows evidence of dispersion, with longer periods arriving first. Love wave dispersion is used to constrain the vertical variations in shear wave velocity in the crust and upper mantle along the travel path between source and receiver. The instantaneous amplitude, the top curve, is a portion of the instantaneous amplitude curve shown in Figure 4.3. The instantaneous frequency, the bottom curve, is proportional to the time rate of change of the phase of the analytic signal when expressed in polar form. The frequency units are mHz, and the instantaneous-frequency graph quantifies the increasing frequency of the dispersed Love waves; the plot is offset by −150. The instantaneous-frequency value at the left intercept is 28.5 mHz, in the middle about 47 mHz, and, at the right, 53.4 mHz, corresponding to periods of 35.1, 21.3, and 18.7 seconds. These periods are approximately what one would measure from the times between the time series peaks. For example, the time between the second and third time series peaks is about 23 seconds.

frequencies. The group velocity curves are used to estimate the crust and upper mantle shear wave velocity structure along the travel path, which in this case is largely oceanic.

4.10 Chapter Summary

To make the transition from the Fourier series to the Discrete Fourier Transform (DFT) we first developed the Fourier series in terms of complex sinusoids, using them to replace real-valued sines and cosines. The DFT is then seen as a Fourier series truncated at the Nyquist frequency, as appropriate for the sampling theorem. When computing and making use of the DFT (in the frequency domain) the main points are as follows.

- The Fourier frequency values for a time series of length N are at integer multiples of one cycle per record length, $1/N$, corresponding to a frequency in Hz of $1/(N\Delta t)$ if the sample interval Δt is specified.
- Both the forward and inverse DFTs are identical in form, and are computable using the same Fast Fourier Transform algorithm.
- In the frequency domain array of sinusoidal coefficients, the negative-frequency coefficients appear in reverse order after the positive-frequency coefficients. If negative times are needed when computing the DFT of a sampled even or odd function in the time domain, they may be interpreted as being in the reverse order after the positive times.
- For real-valued time series, the DFT values have Hermitian symmetry, the positive- and negative-frequency values being complex conjugates of one another.
- The most common normalization convention is to scale the inverse transform by $1/N$ (reciprocal of the record length).
- The cosine and sine coefficients of a real-valued Fourier series representation of a time series may be found from the real and imaginary parts of the corresponding complex DFT values, after proper scaling (multiplying by $2/N$ with standard normalization).
- While it is useful to develop the DFT as a Fourier series, it has other interpretations, both as an evaluation of the Z transform of a time series, and as a length-preserving transformation between complex-valued vector spaces.
- The DFT has many applications, some presented in this chapter (interpolation, analytic signal computation), others in later chapters (linear filtering, transfer function computation, spectral analysis), and many more beyond these.

Exercises

4.1 **Illustrating DFT Properties.** Use a time series with the number of samples $N = 32$. Compute the following and/or illustrate the stated property of the DFT. The time series values may be random numbers or a computed sinusoid, as appropriate.

A. Show that computation of the DFT, followed by computation of its inverse, obtains the original time series, within numerical precision.

B. Find arrays containing the 32 frequencies of the DFT for the cases when $\Delta t = 1$ (dimensionless), 0.002 seconds, 1 day, 200 years. Give appropriate units for these frequencies as well as numerical values.

C. Demonstrate that, when your time series consists of samples of a single Fourier frequency, so that it can be written as $c\cos(2\pi ft) + s\sin(2\pi ft)$ (c and s are numbers of your choosing), all values of the DFT are zero except at that particular (positive and negative) frequency. Show that the Fourier series coefficients (the cosine and sine coefficients) are found from the real and imaginary parts of the DFT after correction for the time series length.

D. Construct a 32-sample time series which is odd, so that its DFT is purely imaginary. Construct a time series which is even, so that its DFT is purely real. Then use the DFT to find the even and odd parts of your original random number series from part A.

E. Show that when x consists of samples of a pure sinusoid as before, but at a frequency that is not one of the Fourier frequencies, then the values of X are largest at the nearest frequencies, but are non-zero at almost all frequencies. Explain why they are non-zero.

4.2 **Time Domain Interpolation with the DFT.** Padding zeros in the frequency domain interpolates in the time domain. but it must be done with care to preserve Hermitian symmetry. For example, a time series $[x_0, x_1, x_2, x_3]$ of length 4, at times $[0, 1, 2, 3]$ has DFT $[X_0, X_1, X_2, X_3]$. Midpoint interpolation creates an eight-point DFT that has the following values: $Xpad = [X_0, X_1, 0.5 \times X_2, 0, 0, 0.5 \times X_2, X_3]$. More zeros inserted between the two Nyquist values interpolates more finely in time. Inverse transforming (assuming $1/N$ normalization on the inverse transform) requires multiplying by 2 in this case.

For the time series $x = [0, 1, 2, \ldots, 31]$, interpolate by a factor 10, so the new time series sampling interval changes from 1 to 0.1. Plot the original 32 points using a plotting symbol that allows them to be seen when plotted with the interpolated values. This is an approximate replication of Figure 4.2.

4.3 **DFT Filtering.** This exercise makes use of the DFT to smooth (that is, low-pass filter) a climate time series. Simple low-pass filtering is accomplished by computing the DFT, setting the frequency values to zero at high frequencies, while preserving Hermitian symmetry, then inverse transforming. Further discussion of DFT filtering appears in Chapter 6.

Use the DFT to low-pass filter the multivariate ENSO index time series to retain frequencies below 0.9 cycles per year. Plot the filtered and original series together, offset. You will need to determine the frequency values associated with the DFT for this series by determining the time series length and the time interval between samples.

5 Linear Systems and Digital Filters

A filter may be a physical system or computational algorithm with an input and output. If the filter is linear, then the relationship between input and output is the same regardless of the amplitude of the input. Throughout this chapter and the entire book we consider only time-invariant linear filters, that is, linear filters whose properties do not change with time. As a consequence, linear systems and filters obey a superposition principle, so that when two inputs are added together the output is the sum of the separate outputs that would result from separate inputs. Another consequence is that a single-frequency sinusoidal input produces a sinusoidal output at exactly the same frequency and no other. As a result, linear systems and filters are preferred models for physical processes and for data processing because they allow analysis and implementation in both the frequency and time domains. The DFT presented in the previous chapter is the main tool for frequency domain analysis and implementation. This chapter develops important elements used in time domain implementation: digital filter equations and discrete convolution; the transfer function; and the impulse response. These concepts are extended to the properties of a cascade of linear filters (the successive application of several different filters to a time series) and are used to define the concept of an inverse filter. Example applications of linear filters in data processing, as models of physical processes, and in methods for finding practical inverse filters appear in Chapters 8 and 9.

5.1 Linear Filter Equations

We begin by presenting filter equations for the two general types of linear filters, moving average (MA) and autoregressive moving average (ARMA) filters. The simplest type of linear filter is the MA filter, which computes, as its output, a linear combination of input values. If x_t is the input and y_t the output, a simple example is

$$y_t = b_{-1}x_{t+1} + b_0 x_{t-0} + b_1 x_{t-1}$$

where $[b_{-1}, b_0, b_1]$ are a set of numbers known as filter coefficients Filter coefficients may take on any values but only in the special case $[1/3, 1/3, 1/3]$ is the output an actual average of the input values. In this example, y_t is computed from x_t at times $t + 1$, t, and $t - 1$, so it requires the input at the future time $t + 1$. This makes it acausal. A causal filter requires input values only at present and past times. The distinction is important in real-time applications, but both causal and acausal filters may be used with recorded data, where past and future samples are both available. The filter coefficient subscripts denote the shift (lag) of the input values they multiply, relative to the output. Therefore b_1 is the coefficient that multiplies x_{t-1}, the input lagged by 1 time unit, b_0 is the coefficient of x_t, the input lagged by 0 time units, and b_{-1} is the coefficient of x_{t+1} that leads by (or has negative lag of)

1 time unit. We will work only with causal filters, however, for which an equivalent filter with 3 coefficients would be

$$y_t = b_0 x_{t-0} + b_1 x_{t-1} + b_2 x_{t-2}$$

For causal filters, the filter order is the largest lag, so this is an MA filter of order 2.

A more complicated type of linear filter equation includes feedback terms, where linear combinations of past outputs are combined with MA terms. Filters containing feedback terms are autoregressive (AR), for example

$$a_0 y_t = b_0 x_t - a_1 y_{t-1}$$

where we have introduced a coefficient a_0 for y_t. The coefficient a_0 is needed for consistent terminology in evaluating transfer functions, to be discussed shortly. However, in all cases $a_0 = 1$. This is an AR filter of order 1, and the MA order is 0. The negative sign for AR coefficients $(-a_1)$ is conventional. A general autoregressive moving average (ARMA) linear filter of order (m, n) combines up to m past output values with up to n past input values:

$$a_0 y_t = b_0 x_t + b_1 x_{t-1} + \cdots + b_n x_{t-n} - a_1 y_{t-1} - \cdots - a_m y_{t-m} \qquad (5.1)$$

A pure AR filter is of order $[m, 0]$, and a pure MA filter is of order $[0, n]$.

Filtering a time series x_t with a pure MA filter is accomplished using a discrete convolution algorithm, as will be described in the next two sections. An algorithm to implement an AR or ARMA filter is more complicated because the feedback terms contribute linear combinations of prior values of the filter output.

5.2 Discrete Convolution

The scaling, shifting, and addition appearing in digital filter equations is called discrete convolution and is denoted by the symbol $*$. (The closely related operation of correlation is denoted by the symbol $\star$.)

Using the MA filter above as an example, the MA filter $y_t = b_0 x_t + b_1 x_{t-1} + b_2 x_{t-2}$ can be written as $y_t = b_t * x_t$. Discrete convolution is visualized as a mechanical operation involving shifting, multiplication, and addition of time series values written on paper. The exercise starts by writing down one series on a sheet of paper, with values separated by commas. Next, on a separate strip of paper, the other series is written in reverse order, with the same space between the values as on the paper sheet. Let x_t be the series on the sheet of paper, and b_t those on the paper strip, in reverse order. We can verify that convolution is commutative, so the the same result is obtained if b_t is written on the sheet and x_t in reverse order on the paper strip.

To calculate the output y_t at each time we form a sum of products of values of b_t multiplying values of x_t adjacent to them. Term by term multiplication and addition of the products gives the convolution result for the time corresponding to x_t, where b_0 is the multiplying coefficient. Then the strip of paper moves one place to the right and the operation is repeated. This generates successive values of the convolution result, in this case the time series y_t. The following illustrates this schematically.

$$x_{t-4}, x_{t-3}, x_{t-2}, x_{t-1}, x_t, x_{t+1}, \ldots$$

$$\ldots, [\, b_2, \quad b_1, \quad b_0 \,] \longrightarrow, \ldots$$

Here the brackets around the three non-zero values of the b series define the paper strip, upon which the b series is written in reverse order. The arrow indicates the left to right motion of the paper strip, and the output at time t is

$$y_t = b_0 x_t + b_1 x_{t-1} + b_2 x_{t-2} \tag{5.2}$$

and at the next time, $t + 1$, it is

$$y_{t+1} = b_0 x_{t+1} + b_1 x_t + b_2 x_{t-1}$$

This example involves just a pure MA filter, but the general form of the ARMA filter shows a convolution operation in which all AR terms are placed on the left-hand side, so an ARMA filter equation can be expressed either as

$$a_0 y_t + a_1 y_{t-1} + \cdots + a_m y_{t-m} = b_0 x_t + b_1 x_{t-1} + \cdots + b_n x_{t-n}$$

or as

$$a_t * y_t = b_t * x_t$$

In actual computations all series are of finite length, so the operation of sliding the strip of paper must deal with the endpoints. With the two series taken to be transients (equal to zero outside their finite length), discrete convolution produces another transient series whose length is the sum of the two lengths, less one, because the series are zero outside their finite lengths. For example, for the time series $g_t = [g_0, g_1, g_2]$ and $h_t = [h_0, h_1]$ their convolution is, using the procedure described above,

$$g_t * h_t = [(g_0 h_0), (g_1 h_0 + g_0 h_1), (g_1 h_1 + g_2 h_0), (g_2 h_1)]$$

a new time series of length $2 + 3 - 1 = 4$. If h_t is viewed as the filter, and g_t as input, then the output $g_t * h_t$ has four terms, but only the two middle values involve a full overlap of h_t with g_t. In filtering operations these end terms might be discarded or retained, depending on the situation.

Discrete transient convolution corresponds one-to-one with polynomial multiplication, so transient convolution can also be obtained by multiplying Z transforms of each time series, as we will now show. Recall from the development of the DFT that the Z transform is a polynomial in Z using time series values as coefficients of the variable Z raised to the integer power of the time t. In the above case $G(Z) = g_0 Z^0 + g_1 Z^1 + g_2 Z^2 = g_0 + g_1 Z + g_2 Z^2$ and $H(Z) = h_0 + h_1 Z$. The product of the Z transforms is the polynomial

$$G(Z)H(Z) = (g_0 h_0) + (g_1 h_0 + g_0 h_1)Z + (g_1 h_1 + g_2 h_0)Z^2 + (g_2 h_1)Z^3$$

which is the Z transform of the discrete transient convolution

$$g_t * h_t = [(g_0 h_0), (g_1 h_0 + g_0 h_1), (g_1 h_1 + g_2 h_0), (g_2 h_1)] \tag{5.3}$$

This result is a convolution theorem for the Z transform. It also verifies that discrete convolution has properties in common with ordinary multiplication. Convolution is therefore commutative, associative and distributive over addition. In Chapter 6 we adapt this Z transform convolution theorem to a form of convolution theorem for the DFT.

A computer algorithm for discrete transient convolution can be devised by noting in this example that the sums of the subscripts in each term on the right-hand side correspond to the integer time of the output series value. That is, the sum of subscripts for the first term in $g_t * h_t$ (for $t = 0$) is zero, for the next term ($t = 1$) the subscripts add to one, for the third term ($t = 2$) the subscripts add to two, and for the fourth term ($t = 3$) they add to three. So, a computer code for convolution can be written by defining integers K to count position in the output series, I to count position in g_t, and J to count time in h_t. All values of the output are set to zero, and two nested loops count through the ranges of I and J to calculate the products $g(I)h(J)$, which are accumulated in a sum at output-series array position $K = I + J$.

5.3 Correlation

Correlation is an operation closely related to convolution. We introduced it in Chapter 2 in the form of autocorrelation, one of four time series statistics. By convention the mean value is removed prior to computing a correlation or autocorrelation statistic. The autocorrelation r_τ of h_t is the sum of the lagged products of a time series with other time series values:

$$r_\tau = [h_0 h_1, h_0^2 + h_1^2, h_0 h_1]$$

where the lags $\tau = [-1, 0, 1]$. This is identical to a convolution between $h_t = [h_0, h_1]$ and its time-reversed version $h_{-t} = [h_1, h_0]$. That is,

$$r(\tau) = [h_0 h_1, h_0^2 + h_1^2, h_0 h_1] = h_t * h_{-t} = h_t \star h_t$$

using the symbol $\star$ to denote correlation. The autocorrelation of the time-reversed series h_{-t} is the same as that of h_t, so we conclude that the autocorrelation of a given time series may be the same as that of many other time series. For example, the autocorrelation of the series $g_t * h_t$ is identical to that of $g_{-t} * h_t$, $g_t * h_{-t}$, and $g_{-t} * h_{-t}$. Correlation defined between two different series is cross correlation. For g_t and h_t, for example,

$$g_t \star h_t = [g_2 h_0, g_2 h_1 + g_1 h_0, g_1 h_1 + g_0 h_0, g_0 h_1]$$

Convolution is commutative but correlation is not. The correlation $h_t \star g_t$ is the time-reversed version of $g_t \star h_t$. One application of correlation is in estimating a time shift when x_t and y_t are very similar to one another but misaligned in time. A relatively large value of cross correlation occurs with alignment. The time shift can be found by looking for a correlation maximum. Even if one of the series is corrupted by noise, the correlation statistic is effective for this task. This is the basis for matched and correlation filtering, described in Chapter 9.

If two series x_t and y_t have the same length, the lag $\tau = 0$ occurs when the two series exactly overlap. This zero-lag cross correlation is proportional to the correlation coefficient ρ_{xy} used to measure the similarity between x_t and y_t. Further discussion of this appears in Chapter 10 and Appendix C. The standard deviations $\hat{\sigma}_x$ and $\hat{\sigma}_y$ are used to normalize the magnitude of the correlation coefficient to be less than unity:

$$\rho_{xy} = \frac{r_0}{\hat{\sigma}_x \hat{\sigma}_y}$$

5.4 Convolution Matrices

The convolution of two discrete transient time series may be implemented via matrix multiplication. Each time series can be written as a column vector, so $g_t * h_t$ may be written either as

$$\begin{bmatrix} g_0 & 0 \\ g_1 & g_0 \\ g_2 & g_1 \\ 0 & g_2 \end{bmatrix} \begin{bmatrix} h_0 \\ h_1 \end{bmatrix} = \begin{bmatrix} g_0 h_0 \\ g_1 h_0 + g_0 h_1 \\ g_2 h_0 + g_1 h_1 \\ g_2 h_1 \end{bmatrix}$$

or as

$$\begin{bmatrix} h_0 & 0 & 0 \\ h_1 & h_0 & 0 \\ 0 & h_1 & h_0 \\ 0 & 0 & h_1 \end{bmatrix} \begin{bmatrix} g_0 \\ g_1 \\ g_2 \end{bmatrix} = \begin{bmatrix} g_0 h_0 \\ g_1 h_0 + g_0 h_1 \\ g_2 h_0 + g_1 h_1 \\ g_2 h_1 \end{bmatrix}$$

The convolution matrix in each case contains the time series values in reverse order, shifted one sample to the right on successive rows, just as the paper strip containing the values in reverse order is moved left to right. Convolution matrices play a central role in the development of least squares filters in Chapter 9.

5.5 Transfer Functions

A linear filter transfer function $L(f)$ is a complex-valued function of frequency describing how the amplitude and phase of a sinusoid of frequency f are altered by the filter. As a complex quantity, $L(f)$ is best represented in its polar form. Then the modulus gives the amplification and the exponent gives the phase shift at each frequency. Via a simple example, we will show how to compute the transfer function, and also verify the single-frequency-in–single-frequency-out property of linear filters. This example also shows how complex-valued sinusoids greatly simplify the associated algebra and how the polar form of $L(f)$ is the most useful.

Consider the MA filter given by

$$y_t = (1/4)x_t + (1/2)x_{t-1} + (1/4)x_{t-2}$$

Let the input be the single-frequency sinusoid $x_t = \exp(2\pi i f t)$ sampled at integer times t. The frequency f may be any value but is not specified at this point. We expect that f lies within the Nyquist band, $f = [-1/2, 1/2]$, but if it takes on a value outside this range we would find that the transfer function is periodic in frequency, so that $L(f)$ will be the same for all alias frequencies. Using the filter equation to calculate the output time series:

$$y_t = (1/4)\exp(2\pi i f t) + (1/2)\exp(2\pi i f(t-1)) + (1/4)\exp(2\pi i f(t-2))$$

The properties of exponentials allow the common term $\exp(2\pi i f t)$ to be factored out and the other terms combined. Expressing cosine using complex exponentials, the result, after simplification, is

$$y_t = [1 + \cos(2\pi f)]/2 \times \exp(-2\pi i f) \times \exp(2\pi i f t) = L(f) \times \exp(2\pi i f t)$$

where the transfer function $L(f)$ is the time-independent part of the right-hand side. It equals the output divided by the input $\exp(2\pi i f t)$:

$$L(f) = [\exp(-2\pi i f)] \times [1 + \cos(2\pi f)]/2$$

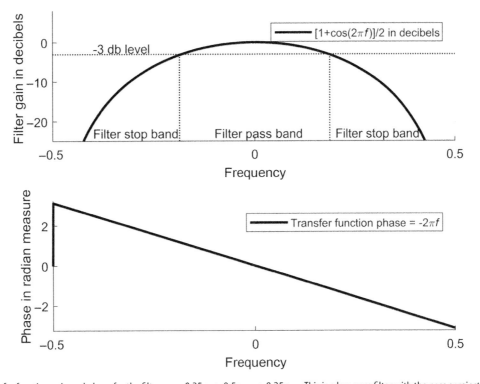

Figure 5.1 Transfer function gain and phase for the filter $y_t = 0.25x_t + 0.5x_{t-1} + 0.25x_{t-2}$. This is a low-pass filter with the pass or reject band defined by the -3 dB level, indicated by the dotted line. Frequencies above the -3 dB level are by convention considered to be passed and those below are in the stop or reject band. The transfer function goes to zero at the Nyquist frequencies, or at negative infinity on a dB scale. The phase is a linear function of frequency. This is due to the time shift of one sample between the causal version of the filter illustrated here, and the acausal version $y_t = 0.25x_{t+1} + 0.5x_t + 0.25x_{t-1}$. The phase of the transfer function for the acausal version is zero at all frequencies. A filter transfer function plot such as this, which shows separate graphs of gain (in dB) and phase (in radians or degrees), is called a Bode plot.

We conclude that the convolution of filter coefficients $[1/4, 1/2, 1/4]$ in the time domain leads to multiplication by $L(f)$ in the frequency domain. In addition, the output is purely sinusoidal, at the same frequency as the input.

When writing $L(f)$ in polar form, its modulus is $[1 + \cos(2\pi f)]/2$, and its phase is $2\pi f$. It is customary to express the modulus in decibels, and to refer to this as the gain. Since both gain and phase are functions of frequency, two graphs are required to illustrate $L(f)$, and are usually plotted together in a Bode plot. The transfer function gain in decibels is $20\log_{10}\|L(f)\| = 10\log_{10}(\|L(f)\|^2)$. The squared modulus of the transfer function, $\|L(f)\|^2$, is known as the power transfer function. If the gain (in dB) is positive then the filter amplifies, and if negative it attenuates. The gain is negative infinity decibels whenever $|L(f)|$ is zero.

Figure 5.1 shows the transfer function of this filter in a Bode plot presenting the gain and phase. The linear change in phase is associated with a time shift; in this case, the time shift is one time sample, associated with the use of a causal filter in place of the zero-phase but acausal (symmetric in time) filter

$$y_t = (1/4)x_{t+1} + (1/2)x_t + (1/4)x_{t-1}$$

The Z transform of the filter coefficient series provides a simple description of the transfer function. For MA coefficients $b_t = [b_0, b_1, b_2] = [1/4, 1/2, 1/4]$ we have

$$B(Z) = (1/4) + (1/2)Z + (1/4)Z^2$$

and

$$L(f) = B(Z)|_{Z=\exp(-2\pi i f)} = Z[1 + \cos(2\pi f)]/2$$

when $Z = \exp(-2\pi i f)$. Note that the transfer function of the acausal version of this filter is

$$(1/4)Z^{-1} + (1/2) + (1/4)Z^1 = [1 + \cos(2\pi f)]/2$$

The causal transfer function differs from the acausal by the multiplicative factor Z. Because there is a one-sample delay between the causal and acausal filters, we can say that multiplication by Z is a one-sample delay operator, and multiplication by Z^n is the delay operator for n samples. These powers of Z cause the phase graph in the Bode plot to have a linear trend, so whenever we see a linear feature in the phase graph, we can assume it is due to a time shift.

For a general ARMA filter, Z transforms of the AR and MA coefficient series, respectively

$$A(Z) = [a_0 + a_1 Z + a_2 Z^2 + \cdots]$$

and

$$B(Z) = [b_0 + b_1 Z + b_2 Z^2 + \cdots]$$

are used to obtain

$$L(f) = \left. \frac{B(Z)}{A(Z)} \right|_{Z=\exp(-2\pi i f)}$$

where it is understood that the ratio is computed separately for each frequency.

The DFT can be used to compute the transfer function using zero-padding to set the number of frequencies. An example follows. Typically, we might compute the transfer function at several hundred frequencies in order to plot it. For example, for an ARMA filter of order (3, 2), $a_t = [1, a_1, a_2, a_3]$ and $b_t = [b_0, b_1, b_2]$. The transfer function is computed at 1000 positive and negative frequencies by padding to length 1000, so

$$L = \frac{DFT(pad_{997}(b_t))}{DFT(pad_{996}(a_t))}$$

where the fraction is evaluated separately at each frequency. The first 501 terms in the array L will be the positive-frequency transfer function over the range $f = [0, f_{Nyquist}]$, and the remaining 499 terms will be the negative-frequency values (in reverse order). The negative-frequency values are complex conjugates of the positive-frequency values when the filter coefficients are real-valued, so are often not plotted.

5.6 Impulse Response

In addition to its transfer function, a digital filter is characterized by its impulse response or IR. The IR is the output time series l_t when the input is $\delta_t = [1, 0, 0, \ldots]$. The quantity δ_t serves the role of a time domain test input, just as a single-frequency sinusoid $\exp(i2\pi ft)$ serves as a test input to find the transfer function; δ_t is an informative test input because it contains all frequencies in equal amounts; its DFT is unity at all frequencies.

For a pure MA filter, the IR is precisely the sequence of filter coefficients, $b_t = [b_0, b_1, \ldots] = l_t$, so for the filter

$$y_t = (1/4)x_t + (1/2)x_{t-1} + (1/4)x_{t-2}$$

the IR is the time series $l_t = [1/4, 1/2, 1/4, 0, 0, \ldots]$. Any MA filter (with a finite number of coefficients) will produce a similar finite-length IR, so all MA filters are finite impulse response or FIR filters. In contrast, AR and ARMA filters are infinite impulse response or IIR filters. For example $y_t = x_t + y_{t-1}$ has $l_t = [1, 1, 1, 1, 1, \ldots]$, which has infinite length. It also never diminishes in amplitude, so the filter is said to be metastable. The IR of a stable filter decays with time, while that of an unstable filter, such as $y_t = x_t + 2y_{t-1}$, with IR $[1, 2, 4, 8, \ldots]$ grows without bound. All MA filters with a finite number of coefficients are stable, and any useful filter will be stable.

The impulse response can be used to calculate the output for any input, even if we do not know the digital filter equation, so the impulse response plays the same role as does a Green's function for systems described by a linear differential equation. For any input, the output is the convolution between the input and the impulse response l_t:

$$y_t = l_t * x_t \tag{5.4}$$

This is evident for an MA filter because $l_t = b_t$, but is also true for AR or ARMA filters. Linearity and superposition properties can be used to understand this. Any input series $x_t = [x_0, x_1, \ldots]$ may be represented as a sum of scaled and delayed δ_t sequences:

$$x_t = x_0\delta_t + x_1\delta_{t-1} + x_2\delta_{t-2} + \cdots = x_t * \delta_t$$

Because l_t does not change with time, the output is the sum of outputs from the separate scaled and shifted impulses, which are just scaled and shifted impulse response time series; thus the sum is

$$y_t = x_0 l_t + x_1 l_{t-1} + x_2 l_{t-2} + \cdots$$

The result is recognized as the convolution $l_t * x_t$.

5.7 Filter Cascades and Inverses

A filter cascade is formed by the application of several linear filters in succession. For a cascade of two filters, the output from Filter 1 becomes the input to Filter 2, and so on. If Filter 1 has impulse response $(l_1)_t$ and transfer function $L_1(f)$ and Filter 2 has impulse response $(l_2)_t$ and transfer function $L_2(f)$ then the filter cascade corresponds to a single filter whose impulse response is the convolution of the two impulse responses, $(l_1)_t * (l_2)_t$, and whose transfer function is the product of the two transfer functions, $L_1(f)L_2(f)$. This extends to cascades of many filters, and we see that the order of applying linear filters does not matter, because both convolution and multiplication are commutative. For example, if ARMA Filters 1 and 2 have transfer functions

$$L_1 = B_1(Z)/A_1(Z)$$

and

$$L_2 = B_2(Z)/A_2(Z)$$

then their cascade has transfer function

$$L(Z) = L_1 L_2 = \frac{B(Z)}{A(Z)} = \frac{B_1(Z)B_2(Z)}{A_1(Z)A_2(Z)} \tag{5.5}$$

In Section 5.2 an example demonstrated that discrete transient convolution corresponds one-to-one with polynomial multiplication. Thus the AR filter coefficients of the cascade are the discrete convolution of the AR coefficients of Filters 1 and 2, and similarly for the MA coefficients. Furthermore, the polynomials $A(Z)$ and $B(Z)$ can be factored and therefore represented as a cascade of many filters. This idea is developed further in Chapter 8 on linear filter design.

The concept of an inverse filter is familiar from home music systems, where poor loudspeakers (with only a limited ability to produce low-frequency sound) and acoustically muffled rooms (with carpets and drapes attenuating the high frequencies) filter the sound. Approximate inverses of loudspeaker filters and muffled room filters are available using the tone controls on most stereo amplifiers. These controls are adjusted to boost the low and high frequencies. You adjust the tone control so that the cascade of tone control, loudspeakers, and room acoustics yields an approximately constant transfer function over the range of frequencies in the music. The tone control is then an inverse filter for cheap loudspeakers and poor room acoustics.

The inverse to a filter with impulse response l_t has impulse response l_t^{-1}. Here the superscript -1 means the convolutional inverse. To understand how a filter and its inverse are related, suppose that time series y_t and x_t are related by a linear filter, $y_t = l_t * x_t$. When l_t^{-1} is convolved with y_t the result is x_t:

$$x_t = l_t^{-1} * y_t$$

The convolution of l_t and l_t^{-1} must be related by

$$l_t * l_t^{-1} = \delta_t = [1, 0, 0, \dots]$$

and the transfer function of the inverse filter must be $1/L(f)$. The product of the two transfer functions is unity at all frequencies. For an ARMA filter whose transfer function is $B(Z)/A(Z)$, its inverse has transfer function $A(Z)/B(Z)$, so the digital filter equation of the inverse is found by interchanging the MA and AR coefficients (Section 5.2). Of course, exact recovery of x_t will be impossible if the transfer function $L(f)$ is zero at any frequency because y_t would not contain information about x_t at such missing frequencies. In this case, and in most practical cases, an approximate inverse filter is required. In Chapter 9 we show how to develop approximate inverses using least squares.

5.8 Chapter Summary

Linear filters are the principal tool for the processing and analysis of time series, and they are the first choice as models for physical systems; examples and applications are given in Chapter 8. Our purpose here has been to develop the important linear filter descriptors, including the digital filter equation, the transfer function, and the impulse response. The main points are as follows.

- A digital filter equation is defined by an array of MA coefficients (the b array), and AR coefficients (the a array).
- The impulse response of a filter l_t is the output when the input is the sequence ($\delta_t = [1, 0, 0, 0, \ldots]$).
- For any linear filter, the output may be computed either by the filter equation, which may involve feedback (autoregressive) terms, or by convolution with the impulse response.
- For purely MA filters, the impulse response is the b array (5.2) considered as a time series.
- A cascade of linear filters has a transfer function equal to the product of individual transfer functions. The cascade impulse response is the convolution of the impulse responses of the individual filters. The cascade filter MA coefficients are the convolution of the b arrays of the individual filters in the cascade. Similarly, the cascade filter AR coefficients are the convolution of the a arrays of the individual filters in the cascade.
- The Z transform polynomials of the a and b arrays, $A(Z)$ and $B(Z)$, respectively, provide a concise way to represent the transfer function, as $L(f) = B(Z)/A(Z)$ where $Z = \exp(-2\pi i f)$, where f is a particular frequency in the Nyquist band.
- The DFT can be used to compute transfer functions, by padding the a and b arrays with zeros.
- For a complex sinusoidal input $\exp(i 2\pi f t)$, the output of a linear filter is given by $L(f)\exp(i 2\pi f t)$. Therefore, the transfer function at any frequency is equal to the output at that frequency divided by the input.
- The (usually) complex-valued transfer function is best represented in polar form, as $|L(f)|\exp(i\phi(f))$. It is customary to plot separate graphs (Bode plots) of the gain (in decibels as $20\log_{10}(|L(f)|)$ and the phase $\phi(f)$.
- The inverse to a filter is formally defined to have a reciprocal transfer function $1/L(f)$, so that when a filter and its inverse are applied in a cascade, the resulting transfer function is unity at all frequencies. In practice, a useful inverse filter may not exist, for example if $L(f)$ is zero at some frequency. Chapter 9 deals with methods for finding practical inverses in such cases.

Exercises

5.1   **Transfer Function Calculation.** Write a function that calculates the transfer function of a general ARMA filter using the ratio of the DFT of the MA coefficients (b_t) and the DFT of the AR coefficients (a_t), using zero-padding as described in this chapter.

5.2   **Inverse Filters.** A particular AR filter has coefficients $a = [1, -0.2, 0.6]$ and $b = [1]$.

   A. Find the coefficients of the inverse filter and verify computationally that the convolution of the impulse responses of the filter and its inverse is δ_t. Because the IR of one of the filters is infinite in length, your convolution may contain some stray non-zero terms because it is done with finite-length time series.

   B. Also verify that the product of the transfer functions of the filter and its inverse is unity at all frequencies.

   C. Plot the gains of the filter and its inverse on a dB scale to show that in decibels they are additive inverses.

   D. Find an order-18 MA filter that has approximately the same transfer function as the original filter. Plot the transfer function of this MA filter together with that of the original AR filter to confirm this.

5.3   **Filter Cascades.** An ARMA filter that passes frequencies below three-quarters of the Nyquist frequency has coefficients

$$b = [0.3468, 1.3873, 2.0809, 1.3873, 0.3468]$$

and

$$a = [1.0000, 1.9684, 1.7359, 0.7245, 0.1204]$$

An ARMA filter that passes frequencies above one-quarter of the Nyquist frequency has coefficients

$$b = [0.4459, -1.3377, 1.3377, -0.4459]$$

and

$$a = [1.0000, -1.4590, 0.9104, -0.1978]$$

Find a single ARMA filter (a band-pass filter) that passes frequencies between one-quarter and three-quarters of the Nyquist. Plot the filter gain on a dB scale to verify that it is flat in the pass band and that at the cutoff frequencies (one-quarter and three-quarters Nyquist), the transfer function is 3 dB below its peak.

6 Convolution and Related Theorems

This chapter is concerned with the relationship between frequency domain operations (using the DFT from Chapter 4) and time domain operations (using discrete transient convolution from Chapter 5). Frequency and time domain operations are connected by a convolution theorem, showing that multiplication in one domain corresponds to convolution in the other. This provides guidance to avoid unwanted wrap-around effects when performing DFT linear filtering. It also motivates the use of window functions as multipliers in both the time and frequency domains to improve linear filtering performance and DFT spectral estimates, as described in Chapter 10.

6.1 Convolution Theorem for the Z Transform

In Section 5.2 it was demonstrated that discrete transient convolution corresponds one-to-one with polynomial multiplication. As a result, the Z transform of the convolution of two time series $g_t * h_t$ is equal to the product of their Z transforms, $G(Z)H(Z)$. This is the convolution theorem for Z transforms, and it can be restated, in a modified fashion, using the DFT. This is so because the DFT is an evaluation of the Z transform of a time series for particular values of $Z = \exp(i2\pi f)$, where f is in the set of Fourier frequencies. Once we understand how to do this, we can perform discrete transient convolution using the DFT. This is an alternative to the sliding, multiplication, and addition algorithm described in Chapter 5. Somewhat surprisingly, performing discrete convolution using the DFT may be computationally more efficient owing to the speed of FFT algorithms.

To adapt the Z transform convolution theorem to the DFT we must recognize that Fourier frequencies are determined by record length. Two different time series may have different record lengths, and their transient convolution will have yet another length. For the Z transform convolution theorem to apply to the DFT, all series must be given the same length by zero-padding. For example, with g_t of length 3 and h_t of length 2, their discrete convolution will be of length 4 (see Section 5.2), so the minimum padding required is one zero onto g_t and two zeros onto h_t. Any additional number of zeros may be padded onto all three series provided they have the same length. Because the number of zeros to be padded is not set, we use the qualifier $pad()$ to indicate the padded version of a time series, without specifying the exact number of zeros. Then the convolution theorem for the DFT, true at each Fourier frequency, is

$$DFT(g_t * h_t) = DFT(pad(g_t))DFT(pad(h_t))$$

This allows the discrete convolution of two series to be performed by first zero-padding, then computing the DFT of each series, multiplying the values at each frequency, and finally computing the IDFT of the product.

6.2 DFT Circular Convolution Theorem

A common application of the DFT is to perform linear filtering of a time series x_t. Suppose that a transfer function $L(f)$ is sampled at the Fourier frequencies corresponding to DFT values X_m and determined by the record length of x_t. The transfer function samples L_m multiply X_m at each Fourier frequency, and the filtered time series is the IDFT of the product. A DFT convolution theorem describes how this frequency domain multiplication corresponds to a new form of circular convolution in the time domain. Circular convolution is illustrated by, for example, a length-4 time series,

$$x = [x_0, x_1, x_2, x_3]$$

with associated DFT values

$$X = [X_0, X_1, X_2, X_3]$$

The transfer function is sampled at the same four Fourier frequencies, giving

$$L = [L_0, L_1, L_2, L_3]$$

The sampled values of the transfer function in the array L will have Hermitian symmetry in the usual case of the filtering of real-valued time series, so its associated time domain series l will be real-valued:

$$l = IDFT(L) = [l_0, l_1, l_2, l_3]$$

The filtered time series is the inverse transform of a time series formed from term-by-term multiplication of the sampled transfer function by the above DFT:

$$IDFT([L_0X_0, L_1X_1, L_2X_2, L_3X_3])$$

which is a time series with four terms:

$$[(l_0x_0 + l_3x_1 + l_2x_2 + l_1x_3), (l_1x_0 + l_0x_1 + l_3x_2 + l_2x_3),$$
$$(l_2x_0 + l_1x_2 + l_0x_3), (l_3x_0 + l_2x_1 + l_1x_2 + l_0x_3)]$$

Each term in this circular convolution is the sum of products of elements of the two time domain series, as in ordinary convolution.

Circular convolution can be demonstrated by the mechanical operation of writing down one series in reverse order on a strip of paper and sliding it past the other, forming the sum of the term-by-term products, as for transient convolution. However, the series written down in its usual order is supplemented by placing a copy of it both on the left and on the right, as illustrated below. There are four terms that overlap the four terms of the series written in reverse order; these are multiplied and added to obtain the first term in the circular convolution. To find the second term in the circular convolution, move the second row to the right by one position, filling in an additional zero on the left and removing one on the right. Repeating this two more times yields the third and fourth terms of the circular convolution. One more shift of the second row creates a repeated first term, and the convolution is complete:

$$\begin{array}{cccccccccccc} x_0 & x_1 & x_2 & x_3 & x_0 & x_1 & x_2 & x_3 & x_0 & x_1 & x_2 & x_3 \\ 0 & l_3 & l_2 & l_1 & l_0 & 0 & 0 & 0 & 0 & 0 & 0 & 0 \end{array}$$

Thus the convolution theorem can be written

$$x_t \odot l_t = IDFT(LX) \tag{6.1}$$

where $LX = [L_0 X_0, L_1 X_1, L_2 X_2, L_3 X_3]$ and $\odot$ means circular (also called cyclic) convolution. Matrix multiplication can also be used to describe circular convolution where the roles of l_t and x_t may be interchanged; the elements of the column vector resulting from the matrix multiplication below give the circular convolution terms in (6.1)

$$\begin{bmatrix} x_0 & x_3 & x_2 & x_1 \\ x_1 & x_0 & x_3 & x_2 \\ x_2 & x_1 & x_0 & x_3 \\ x_3 & x_2 & x_1 & x_0 \end{bmatrix} \begin{bmatrix} l_0 \\ l_1 \\ l_2 \\ l_3 \end{bmatrix}$$

The circular convolution theorem is essential to understanding DFT linear filtering, since DFT filtering is easy to implement and computationally efficient but inevitably creates a filtered time series that mixes values from the beginning and end of the series. This is called wrap-around. Likewise, circular convolution appears in the use of window functions to minimize the smearing of power spectrum peaks in spectral analysis. Both are discussed in more detail below.

6.3 Autocorrelation Theorem

We now demonstrate by example that the DFT of the autocorrelation of a time series is equal to the squared modulus of its DFT, after proper zero-padding. This is an autocorrelation theorem for discrete time series and is analogous to a similar result for the Fourier transform of continuous functions, outlined in Appendix B. The squared modulus of the DFT is the basis for periodogram power spectrum estimates (describing how variance is distributed over frequency), developed in Chapter 10. Therefore, both the autocorrelation and the power spectrum contain the same information.

For a zero-mean time series $h_t = [h_0, h_1, h_2]$ (as an example), the autocorrelation is (see Section 5.3)

$$r_\tau = [h_0 h_2, h_0 h_1 + h_1 h_2, h_0^2 + h_1^2 + h_2^2, h_0 h_1 + h_1 h_2, h_0 h_2] = [r_{-2}, r_{-1}, r_0, r_1, r_2]$$

Here the lag τ values are given by the subscripts in the final expression. They are symmetric about $\tau = 0$. The autocorrelation can also be computed as the convolution with its time-reversed self:

$$r_\tau = h_t \star h_t = h_t * h_{-t}$$

The Z transform of the autocorrelation, $R(Z)$, is the product of the Z transforms of the time series and its time-reversed version:

$$R(Z) = [h_0 + h_1 Z^1 + h_2 Z^2][h_0 + h_1 Z^{-1} + h_2 Z^{-2}]$$

After multiplying and gathering common powers of Z, the result has equal coefficients for positive and negative powers of Z, which are complex conjugates of each other, so $R(Z)$ is real-valued. It is also positive:

$$R(Z) = (h_0 h_2)Z^{-2} + (h_0 h_1 + h_1 h_2)Z^{-1} + (h_0^2 + h_1^2 + h_2^2) + (h_0 h_1 + h_1 h_2)Z^1 + (h_0 h_2)Z^2$$

The Z transform $R(Z)$ can be evaluated using the DFT by placing the negative lags in the autocorrelation in reverse order after the positive lags: $r_\tau = [r_0, r_1, r_2, r_{-2}, r_{-1}]$. To compare with the DFT of h_t, we must pad two zeros to make it of length 5. Then the following is true:

$$|DFT(pad_2(h_t))|^2 = DFT(r_\tau)$$

Thus, the two arrays are equal at each frequency. The autocorrelation theorem applies also to complex-valued time series, in which case the autocorrelation is equal to the convolution of a series with its time-reversed complex conjugate, giving the autocorrelation sequence Hermitian symmetry and making its DFT positive and real-valued.

A related implication is that the DFT of a time-reversed series is the complex conjugate of the original. This can be seen from the DFT definition:

$$X_m = \sum_{t=0}^{N-1} x_t \exp(-2\pi i (m/N) t)$$

If the quantity t is replaced by $-t$, the effect is the same as changing the sign of i on the right-hand side, which creates the complex conjugate. We can conclude in general that the operation of correlation in the time domain corresponds in the frequency domain to multiplication by the complex conjugate. If g_t and h_t are two time series then

$$DFT(h_t \star g_t) = DFT(pad(h_t)) DFT(pad(g_t))^* = H * G^*$$

where the left-hand side is the DFT of the cross correlation of h_t with g_t and the right-hand side indicates that padding is required to make all series the same length. The equation is a statement for each frequency at which the DFT is computed.

6.4 Window Functions

Window functions are used to modify a time series, or its DFT, as described below. Windowing can be performed in the time or frequency domains, and application in one domain corresponds to convolution in the other. Windows applied to the time domain are used to improve power spectrum estimates, a topic in Chapter 10. Windows applied in the frequency domain can improve the performance of linear filtering using the DFT, as discussed in the next section.

To understand the role of windows we consider three types among many in the literature. The boxcar, Bartlett, and Hanning windows are shown in Figure 6.1. Considered as time domain windows, they are symmetric with respect to the time series midpoint. As frequency domain windows they are symmetric with respect to zero frequency. As multipliers in the time domain, their effect is convolution in the frequency domain. In the frequency domain, convolution takes place between the

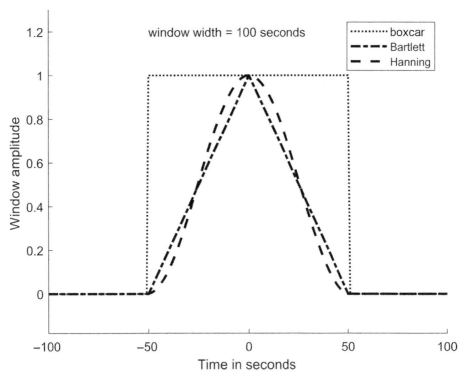

Figure 6.1 Three window functions in common use are the boxcar, Bartlett, and Hanning. These may be used two-sided, as shown, to taper a time series near its endpoints in periodogram spectral computation, as described in Chapter 10. Half the window (one-sided) may be used to taper an impulse response in order to improve digital filter behavior, as described in Chapter 8, or to reduce Gibbs oscillations by modifying a frequency domain transfer function to improve the DFT filtering results. The boxcar is often not considered to be a window because it applies the same weight to all values. The Bartlett is also called a triangular window. The Hanning is one of many types of smooth windows.

time series DFT and the transfer function, as shown in Figure 6.2. This figure shows that the transfer function main central lobes and side lobes are quite different for the three transfer functions. There are relatively large side lobes in the boxcar transfer function, although it has the narrowest central lobe. There are wider central lobes in the Bartlett and Hanning transfer functions but much smaller side lobes.

An important application of window functions occurs in spectral analysis, which separates time series into individual frequency bands. Chapter 10 develops methods for finding power and coherence spectra, which give, respectively, measures of the variance of a time series and of the correlation between two time series, within individual frequency bands. The DFT is the principal tool in spectral analysis. The time series to be analyzed may be viewed as the time domain product of a much longer time series with a boxcar window whose width equals the time series record length. By the convolution theorem, the DFT spectrum values are the convolution of this boxcar transfer function with a true spectrum that might be obtained from the much longer time series. Convolution with the boxcar transfer function, in Figure 6.2, has the effect of smearing or blurring spectrum values in the frequency domain. Both main and side lobes contribute to the blurring associated with

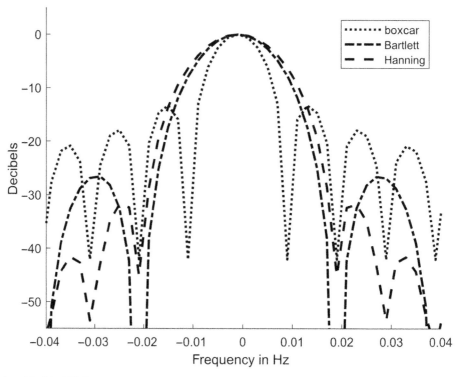

Squared moduli of the DFT of the three windows on a decibel scale. These are frequency domain transfer functions associated with multiplying a time series by one of the three windows in Figure 6.1 of 100-seconds width. The nominal bandwidth (the width of the central lobe) of each is approximately the reciprocal record length, or 0.01 Hz. The boxcar central lobe width (as measured by its −3 dB width) is about 0.01 Hz. The other windows have wider main lobes, associated with somewhat lower frequency resolution, but with much lower side lobes.

convolution, so the relatively large side lobes of the boxcar transfer function are undesirable. If we multiply the time series by either the Bartlett or Hanning windows, the side lobe contributions are diminished, although there is slightly more blurring associated with a wider central lobe. Evidently, there is a trade-off in the use of window functions regarding the main and side lobe contributions to frequency domain blurring. This trade-off has motivated the development of many other windows in traditional spectral analysis, and more than a dozen are in common use.

To provide further intuition concerning window functions, we note that for shorter time series (for which narrower windows are needed) the transfer function will be wider and the blurring worse. For longer time series, the window transfer function will be narrower. These statements are easily verified computationally, and they constitute a property for the DFT that is analogous to the similarity theorem for the Fourier transform of continuous functions described in appendix section B.4. As a result, frequency domain resolution in spectral analysis improves (there is less blurring) for longer time series. Equivalently, as we already know, Fourier frequencies are more closely spaced for longer time series.

The traditional time domain windowing used in spectral analysis has the effect of tapering the time series to zero at its beginning and end points. While this may reduce frequency domain blurring,

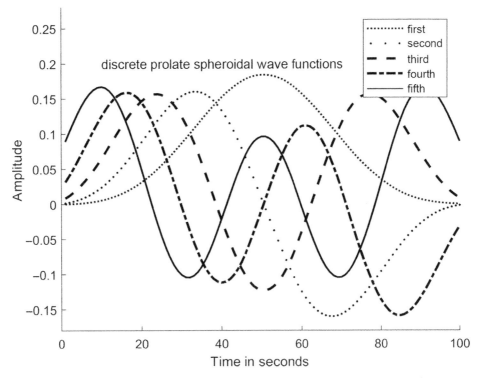

Figure 6.3 The first five discrete prolate spheroidal window (PSWF) functions. The sum of their frequency domain transfer functions is nearly constant over a defined bandwidth. These serve as time domain windows in multi-taper spectral analysis, illustrated in Chapter 10.

depending on the choice of window, it also diminishes the importance of the data near the beginning and end of the time series.

However, the data are presumed to be of the same quality throughout. Thus, in fairness to the data, and for other reasons outlined here and in Chapter 10, an alternative is to use a set of discrete prolate spheroidal wave function (PSWF) windows, also known as Slepian functions. These windows were introduced in the early 1980s. Chapter 10 shows their use in multi-taper method (MTM) power spectrum estimates. Figure 6.3 shows the first five PSWF windows. The first is similar to a Hanning window, but the other four provide weighting to different parts of the time series, each successive window giving more weight toward the ends of the time series. This addresses the concern about using all the data in a fair way to obtain a power spectrum estimate. The PSWF windows have the remarkable property that their frequency domain transfer functions add together to create an effective combined transfer function that is nearly constant over a frequency bandwidth of about 0.05 Hz, as shown in Figure 6.4. This is discussed in Chapter 10 also. In the MTM method, a spectrum (periodogram) is computed separately for each of the five (or more) PSWF windowed versions. Estimates of the spectrum of coherence, also discussed in Chapter 10, are obtained in the same way. The DFT of each windowed version is called an eigenspectrum and the sum of the eigenspectra is the MTM estimate. These estimates reduce the leakage of variance compared with

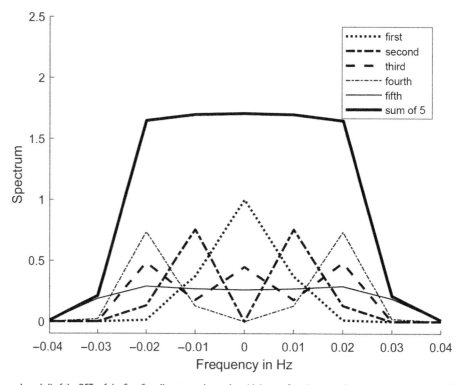

Figure 6.4 The squared moduli of the DFTs of the first five discrete prolate spheroidal wave functions are shown separately, along with their sum. The sum of the five provides a nearly uniform average over a frequency bandwidth of 0.05 Hz for a 100-second time series. The side lobes of the sum are nearly zero, unlike those of a single window shown in Figure 6.2. The low side lobe property of the sum of windows means that the sum of five eigenspectra is an excellent average over this frequency band. An example is presented in Chapter 10.

a standard periodogram, because the effective window (the sum in Figure 6.4) has small side lobes. Figure 10.5 shows the MTM spectral window on a decibel scale.

6.5 Linear Filtering with the DFT

Because FFT algorithms are fast, DFT filtering is often preferred if the computational effort is important. In other cases, the choice of time or frequency domain filtering may be a matter of personal preference. However, when the desired transfer function changes rapidly with respect to frequency (relative to the DFT frequency resolution, equal to the reciprocal of the record length), filtering must be done in the time domain.

The DFT filtering of a time series x is accomplished by computing its DFT, X, multiplying X at each frequency by the array L containing samples of the desired transfer function $L(f)$, and finally computing the IDFT of their frequency domain product. When x is real, both X and L must have Hermitian symmetry. High-pass, low-pass, and band-pass filtering are common applications. Normally we specify the range of high frequencies to be rejected (the stop band) and the range of the frequencies to be retained (the pass band) and, at first glance, the transfer function array L might

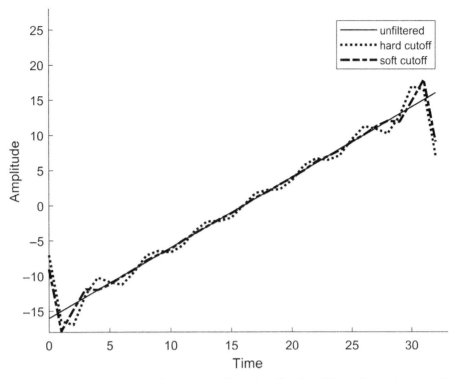

Figure 6.5 A time series ramp function and two low-pass filtered versions. The hard cutoff results in Gibbs oscillations throughout the time series as a result of wrap-around. Using a Bartlett window to create a soft truncation reduces the Gibbs oscillations.

be assigned values of 0 for the stop band frequencies and 1 in the pass band. However, the example in Figures 6.5 and 6.6 shows that it is preferable to use a frequency domain window with a gradual transition (taper or window function) between the pass and stop bands. In addition, if x has a trend or other feature causing a large discontinuity between the first and last values, the trend should be removed prior to DFT filtering. If the trend is to be retained in the filtered series afterwards, it can be added back after the DFT filtering.

To illustrate the value of using a window function as a taper between the stop and pass bands in DFT filtering, consider a time series x of length 33, which is the ramp function in Figure 6.5. This is chosen as a useful example even though the linear trend would not normally be subjected to DFT filtering, as noted. The ramp function discontinuity between the beginning and end values makes clear the effects of wrap-around and Gibbs oscillations.

Suppose that we want to remove frequencies higher than half the Nyquist and to retain lower frequencies in the band $f = [-1/4, 1/4]$. A simple transfer function would correspond to the sharp truncation shown in Figure 6.6. An alternative is a soft linear transition between pass and stop band, effectively a Bartlett window transition. A window with a more gradual transition from pass to stop bands will have lower side lobes. Time domain circular convolution that results from multiplication by the frequency domain transfer function will cause wrap-around and Gibbs oscillations. Figure 6.5 shows that the hard-truncation low-pass filter creates ripples throughout the time series, whereas the soft-transition ripples are confined to the ends, a result of the lower side lobes. The main conclusion from this example is that transfer functions in the frequency domain should have a gradual transition between the stop and pass bands. Additionally, conditioning the time series to avoid discontinuity

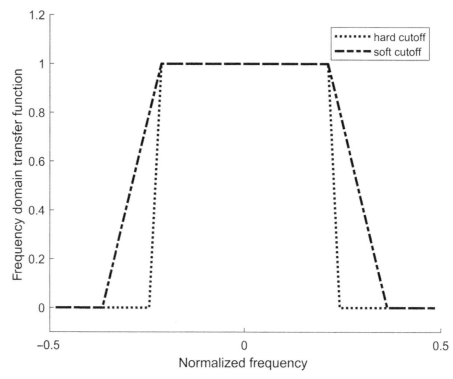

Figure 6.6 Frequency domain transfer functions of the hard and soft (Bartlett window) cutoffs for filtering the ramp function using the DFT.

between first and last samples (intentionally not carried out in this example) is also important. This can involve removing trends and also padding zeros onto the time series.

6.6 Chapter Summary

The convolution theorem and its consequences provide essential knowledge for understanding and improving linear filtering and other operations in both time and frequency domains. The main points are as follows.

- The Z transform convolution theorem shows that discrete transient convolution corresponds one-to-one with Z transform polynomial multiplication, confirming that convolution and multiplication share the same associative, distributive, and commutative properties.
- The Z transform convolution theorem is confirmed using the DFT. This is done by padding zeros to give all the time series the same length, ensuring that the DFT of each series is computed at the same frequencies.
- Frequency domain filtering involves multiplying the DFT of a time series at each frequency by a transfer function array L. The filtered time series is the circular convolution of the original series with the inverse DFT of L. Circular convolution, known as wrap-around, may produce unwanted artifacts in the filtered time series.
- Wrap-around artifacts can be mitigated by conditioning the time series prior to filtering. Conditioning may include zero-padding and the removal of trends and other features which cause

discontinuities between the beginning and end of the series. Another mitigating step is to use window functions as multipliers in the frequency domain to introduce gradual changes in the transfer function array L.

- A consequence of the convolution theorem is that the DFT of the autocorrelation of a time series equals the squared modulus of the DFT of the time series itself. Window functions applied as multipliers in the time domain can be used to improve spectral estimates computed using the DFT. Examples are given in Chapter 10.

- Window functions applied as multipliers in the time domain can be used to modify and improve spectral estimates. Examples are given in Chapter 10.

Exercises

6.1 **Convolution Algorithms and Convolution Matrices.** It is useful to have functions available to perform discrete transient convolution and to be familiar with expressing convolution as matrix multiplication.

A. Write a function that performs discrete transient convolution. The output time series z is the transient convolution of input time series x and y. For example, $[z] = \text{myconv}(x, y)$, where "myconv" means "my convolution". The algorithm in your function for input arrays x and y is described in Chapter 5. First set all z values to zero; then, with two nested loops (integer counters m and n) go through each element of x and y, adding the product $x(m)y(n)$ to z in array position $m + n - 1$.

B. Use your convolution function to verify the properties of discrete transient convolution. Use three time series, each consisting of five simple integers, to verify $a * b = b * a$, the commutative property; $a * (b + c) = a * b + a * c$, the distributive property; $a * b * c = (a * b) * c = a * (b * c)$, the associative property.

C. Construct a convolution matrix for time series a in part B and use matrix multiplication to compute the convolution $a * b$, confirming the same result as the convolution function.

D. Use zero-padding to verify that the Z transform of the convolution of two time series is the product of their Z transforms. Use series a and b from part B.

E. Use the DFT to find the circular convolution of time series b and c from part B.

6.2 **Comparing Linear Filters.** Many different linear filters can be devised to accomplish a given task. The purpose here is to examine the relative performance of two filters designed to remove frequencies above half the Nyquist. Both filters have about the same number of coefficients, so the performance differences will illustrate the advantages of one type over another. You will need to compute transfer functions, using the DFT as described in Chapter 5.

A. An 11-term MA filter for this task is presented in Chapter 8. Here we want to examine the transfer functions in order to become familiar with transfer function descriptive terms, including pass band, stop band, and roll-off (see below), and to understand how to implement a filter in zero-phase form. As will be described in Chapter 8, the coefficients of a pure MA filter are samples of the function $(0.5 \ \text{sinc}(t/2))$ so

$$b = [0.0637, -0.0000, -0.1061, 0.0000, 0.3183, 0.5000,$$
$$0.3183, 0.0000, -0.1061, -0.0000, 0.0637]$$

When modified by multiplying term-by-term with the following Hanning window,

$$[0.0670, 0.2500, 0.5000, 0.7500, 0.9330, 1.0000,$$
$$0.9330, 0.7500, 0.5000, 0.2500, 0.0670]$$

the windowed filter coefficients are

$$b = [0.0043, -0.0000, -0.0531, 0.0000, 0.2970, 0.5000,$$
$$0.2970, 0.0000, -0.0531, -0.0000, 0.0043]$$

An ARMA filter for the task of passing half the Nyquist band is the Butterworth filter of order (4, 4). The number of coefficients in the Butterworth filter is about the same as in the pure MA filter, so the same computational effort is required. The Butterworth coefficients are

$$b = [0.0940, 0.3759, 0.5639, 0.3759, 0.0940]$$

$$a = [1.0000, 0.0000, 0.4860, 0.0000, 0.0177]$$

Plot the gain in decibels for the three filters (truncated sinc samples, Hanning-windowed sinc samples, and Butterworth) on the same graph, for about 500 frequencies in the range zero to Nyquist, with a vertical scale [−60, 10] dB. This will replicate Figure 8.9. Find the cutoff frequency for each filter (that is, the −3 dB point). Which is closer to half Nyquist? A good filter should have a steep slope (called the filter roll-off) at the cutoff frequency. This slope is by convention measured in decibels per octave (an octave interval corresponds to frequency doubling). What are the roll-off slopes for the three filters? Another measure of a good filter is low side lobes in the stop (reject) band. Side lobes are additional peaks that might be below the −3 dB level but are still not desirable. Find the size, in dB below the transfer function peak, of the largest side lobe in the reject band for each filter.

B. Low-pass filters will normally be implemented as zero-phase, because the goal is to remove a range of frequencies but not alter those retained by changing their phase. Chapter 8 describes how to implement any filter to have a phase transfer function equal to zero at all frequencies. This is done by filtering twice, first in the forward and then in the reverse direction. To filter in the reverse direction, simply reverse the order after the first application, then reverse it again after the second pass of the filter. Use a Gaussian zero-mean white noise time series of length 1000 as a test example. Apply the three filters above in zero-phase form (forward and reverse) and plot the smoothed time series, offset above one another, to show that the results are similar. Plot the transfer function gains of the two-pass filters. Find the new cutoff frequencies for each (the −3 dB point) when implemented in zero-phase two-pass form.

7   Least Squares

The least squares method is among the most widely used data analysis and parameter estimation tools. Its development is associated with the work of Gauss, the nineteenth century German geophysicist, who introduced many innovations in computation, geophysics, and mathematics, most of which continue to be in wide use today. We will introduce least squares from two viewpoints, one based on probability arguments and considering the data to be contaminated with random errors and the other based on a linear algebra viewpoint and involving the solution of simultaneous linear equations. We will employ these two viewpoints to develop the least squares approach, using, as an example, the fitting of polynomials to a time series. While this might appear to be a departure from linear filters and related topics, we show in subsequent chapters that in fact least squares serves as a powerful digital filter design tool. We will also find that a stochastic viewpoint (in which time series values are considered to be random variables; see Appendix C) leads also to the use of least squares in the development of prediction and interpolation filters.

7.1   Motivations for Least Squares

The use of least squares can be motivated by recognizing that virtually all observed data contain errors (noise), assumed to be random and describable by a probability density function (pdf) as outlined in Appendix C. While the probability density function of the noise may be unknown, Gauss showed in his central limit theorem (see Appendix C) that normal (Gaussian) behavior is anticipated to hold in many cases. Assuming that the noise behaves as a Gaussian random variable and invoking Bayes' theorem, an extension of the law of compound probability, we can obtain maximum likelihood estimates of the model parameters (such as polynomial coefficients) that are being fitted to the data. If the Gaussian assumption about errors is correct, the estimated parameters, considered as random variables themselves, are least squares estimates. Further discussions of compound probability and Bayes' theorem are given in Appendix C.

From a linear algebra viewpoint the estimation of model parameters (polynomial coefficients, for example) requires solving simultaneous linear equations, especially for the over-determined case when there are more equations than unknowns. An over-determined set of equations generally lacks an exact solution. In that case, a vector of observations (data) is approximated within the vector subspace spanned by columns of the linear equation matrix. The approximation lying within this subspace minimizes the squared distance to the data vector. The main effort in this chapter is to outline the likelihood and the linear algebra development of least squares, to show how modifications can be made for data of variable quality and, after developing the main ideas using the polynomial fitting problem, to review a few other applications.

The development of least squares in this chapter uses only linear problems as examples, those in which the model parameters to be estimated depend linearly on the observed data. However, least squares is commonly used for non-linear problems as well. In that case the problem is usually linearized so that the estimated model parameters are adjustments to the initial model parameters. This approach and other extensions of least squares are important elements of geophysical inverse theory. This chapter provides background material for further study of the latter topic.

7.2 Least Squares via Maximum Likelihood

It is commonly assumed that data errors behave as Gaussian (normally) distributed random variables, because Gauss' central limit theorem predicts that the sum of many random variables (independent of one another but of the same type) will be Gaussian even if the individual random variables in the sum are not. Because we rarely know the precise nature of the errors in data, a Gaussian pdf is the standard assumption in many cases. The Gaussian pdf for random variable n used to describe errors is $p_n(u)$, a function of variable u that can take on any possible value of n:

$$p_n(u) = \frac{1}{\sqrt{2\pi\sigma_n^2}} \exp\left(-\frac{1}{2}[(u - \mu_n)/\sigma_n]^2\right)$$

This is a bell-shaped curve that is symmetric about μ_n, with width determined by σ_n. The area under this curve (or any pdf) is unity. For Gaussian random variables, the area under the curve within the range $[\mu_n - \sigma_n, \mu_n + \sigma_n]$ is about 2/3, where μ_n and σ_n are respectively the mean and standard deviation, so a Gaussian random variable has probability two-thirds of taking on a value within one standard deviation of its mean.

To illustrate the maximum likelihood approach, consider estimating the speed of a vehicle from measurements of position d (contaminated with Gaussian errors) taken at times t, assumed to be free of error. Assuming it is constant, speed m_1 is the slope of a line estimated from N data (t_k, d_k) for $k = [1, 2, \ldots, N]$. Figure 7.1 illustrates a possible set of four data.

The Gaussian random error in each d_k is independent of the others but shares the same standard deviation σ_n. For all the errors, it is assumed that $\mu_n = 0$, so that they have no systematic bias. Thus, we are fitting a polynomial of degree 1 to the (t, d) data, commonly called the regression of d upon t:

$$d_k = m_0 + m_1 t_k + n_k$$

The polynomial coefficients m_0, m_1 constitute the model being fitted to the data. To simplify the problem further, assume for now that $m_0 = 0$ (at time zero, the position of the vehicle is known to be zero).

With only one parameter, the speed m_1, and only one datum ($N = 1$), a single-point slope could be computed:

$$\hat{m}_1 = d_1/t_1$$

where the circumflex $\hat{m}_1$ denotes an estimate. With more data, individual single-point slopes $\hat{m}_1 = d_k/t_k$ for each k could be calculated. Intuition could be used to make use of more data, for example, averaging all single-point slopes, selecting the median single-point slope, or some other choice.

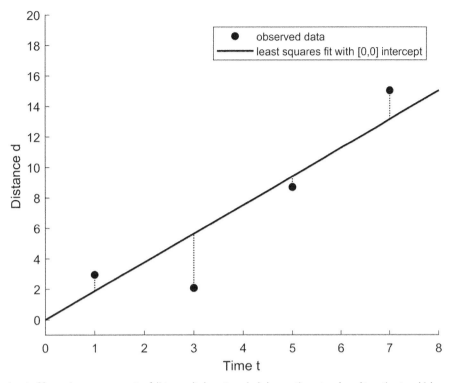

Figure 7.1 An example set of four noisy measurements of distance d taken at precisely known times t, and used to estimate vehicle speed. The vehicle is known to be at distance $d = 0$ at time $t = 0$, so the least-squares-fit line, constrained to pass through $d = 0, t = 0$, is plotted, and its slope is an estimate of the vehicle speed. The least squares solution is that which minimizes the sum of the squared vertical distances (indicated by the dotted lines) to the estimated line. It gives a sensible result, which is easy to obtain by solving linear equations. A different estimate might be obtained using other criteria, for example by minimizing the sum of the absolute values of the vertical distances. Here, the motivation for minimizing the sum of the squared vertical distances is that errors in the four measurements of d are known to be Gaussian (normal) random numbers, each independent of the others and all with the same variance. When this is a correct model for the errors, or any cause of departures from a straight line, the maximum likelihood estimate is the least squares result, as shown in the main text.

Maximum likelihood is another approach, which starts by assuming that, before observations are taken, both the speed m_1 and the data (t, d) are random variables. Then Bayes' theorem (Appendix C) proposes they are related by the law of compound probabilities:

$$P(m_1, data) = P(m_1|data)P(data) = P(data|m_1)P(m_1)$$

where *data* refers in this case to the distance d. The equation reads: "The probability of obtaining both the value m_1 of the speed and at the same time the data values (t, d) equals the probability of the value m_1 of the speed given the data, values (t, d) multiplied by the probability of these data values. It also equals the probability of these data values, given this speed, multiplied by the probability of the speed". Rearranging this gives

$$P(m_1|data) = \frac{P(data|m_1)P(m_1)}{P(data)}$$

The left-hand side is the probability of the speed m_1, given the data (t, d). The right-hand side factor $P(m_1)$ is the prior probability of m_1 without regard to the data. A Bayesian approach takes into account both $P(m_1)$ and $P(data)$, but in ordinary or simple least squares, both the prior probability of m_1 and the probability of the data are taken as constants, so that

$$P(m_1|data) \propto P(data|m_1)$$

The left-hand side of this relation is what we want, but the right-hand side can be determined from the Gaussian assumption about errors and the numerical values of the observed data. When this has been done, we find the value $\hat{m}_1$ that maximizes the right-hand side, called the likelihood of m_1. To see how this works, write down $P(data|m_1)$ for the first data value ($k = 1$, $n_1 = d_1 - m_1 t_1$). The likelihood is

$$\frac{1}{\sqrt{2\pi\sigma_n^2}} \exp\left(-\frac{1}{2}[n_1/\sigma_n]^2\right)$$

Here σ_n is unknown, but is not needed because we are only looking for the value of m_1 that maximizes the likelihood. If $[n_1, n_2, \ldots]$ are independent errors, the probability that all these error values will occur is the product of the individual probabilities, each having the same form as the pdf of n_1. Dropping the multiplying factor containing σ_n, we obtain

$$P(data|m_1)] \propto \prod_1^N \exp\left(-\frac{1}{2}[n_1/\sigma_n]^2\right) \cdots \exp\left(-\frac{1}{2}[n_N/\sigma_n]^2\right)$$

The exponents add in the product, resulting in

$$P(data|m_1) \propto \exp\left(-\frac{1}{2}\sum_{k=1}^N [n_k/\sigma_n]^2\right)$$

The exponent is proportional to the sum of the squares of the n_k and, since the exponent carries a negative sign, the likelihood is maximized when the sum of squares is smallest; σ_n^2 remains unknown but as it is a constant, this does not affect the above conclusion. Therefore the maximum likelihood value for m_1 minimizes the sum of squares

$$\sum_{k=1}^N [n_k]^2 = \sum_{k=1}^N [d_k - \hat{m}_1 t_k]^2$$

The value $\hat{m}_1$ minimizes the sum of squares of the vertical distances between the estimated line $\hat{m}_1 t_k$ and the observed d_k. With N data, the sum of squares $S(m_1)$ can be normalized in various ways, but, when divided by the number of data, N, we obtain

$$S(m_1) = \frac{1}{N}\sum_{k=1}^N [d_k - m_1 t_k]^2 = \frac{1}{N}\sum_{k=0}^N [n_k]^2$$

This quantity should be similar to the error variance for each value of d_k, of which we might have some notion. If $S(\hat{m}_1)$ is much larger than expected individual error variance, the model does not fit the data well. This might indicate that the model is incorrect (for example, the vehicle speed was not constant) or that the errors are larger than expected. Regardless of which explanation applies, it is important to examine model misfit to assess the correctness of the model assumptions.

We can calculate the derivative of S with respect to m_1, and solve for the minimum where this derivative is zero. In this context m_1 is treated as an ordinary variable. The value that minimizes the sum of squares is $\hat{m}_1$, the least squares estimate. This is an extreme value of $S(m_1)$, which must be a minimum rather than a maximum because the right-hand side is always positive. The equation for the estimate is

$$\frac{\partial S(m_1)}{\partial m_1} = 0 = 2 \sum_{k=0}^{N} [d_k - \hat{m}_1 t_k](-t_k)$$

where m_1 has been replaced by $\hat{m}_1$ to indicate that this is the particular value that minimizes the sum of squares. Rearranging yields a "normal" equation

$$\hat{m}_1 \sum_{k=1}^{N} t_k^2 = \sum_{k=1}^{N} t_k d_k$$

leading to the estimate

$$\hat{m}_1 = \frac{\sum_{k=1}^{N} t_k d_k}{\sum_{k=1}^{N} t_k^2}$$

The estimate is linear in d, but involves t in a non-linear way, with the square of t in the denominator. This is still considered a linear estimate because the values of t are presumed to be known perfectly. The least squares solution for the four data in Figure 7.1 shows that the estimate that minimizes the sum of the squared vertical distances to the line appears to be sensible.

To complete the discussion, we can remove the constraint that the intercept m_0 is fixed, and consider fitting both slope and intercept. Then there will be two normal equations from setting to zero the partial derivatives of $S(m_0, m_1)$ with respect to each parameter. Taking derivatives in this case would be tedious and prone to mistakes. If we want to allow for vehicle acceleration, the model would require a third parameter m_2:

$$d_k = m_0 + b_1 t_k + m_2 t_k^2 + n_k$$

This is equivalent to fitting a quadratic polynomial, and would require lengthy evaluation of the partial derivatives of $S(m_0, m_1, m_2)$ with respect to the three parameters, to obtain three normal equations in the three unknowns.

7.3 Least Squares via Linear Algebra

We can represent the above problem of estimating vehicle speed (the regression of d on t) as the solution of a set of over-determined simultaneous linear "observation equations". This approach easily generalizes to more complex models, such as result from the introduction of intercept (m_0) and acceleration (m_2) parameters, and also to yet more complex problems. In general the observation equations have on the left a matrix multiplying a column vector of parameters to be estimated. The left-hand side is a prediction of the data for any values of the parameters. The right-hand side is the column vector of the observed data. The right- and left-hand sides are not exactly equal because of errors in the data. In the first case discussed in Section 7.2, we assumed $m_0 = 0$, so the matrix on the left reduces to a column vector and the column vector of parameters reduces to a scalar. Thus the observation equations on the left-hand side simply reduce to a $1 \times N$ matrix multiplying a column vector with one element:

$$
\begin{bmatrix} t_1 \\ t_2 \\ \ldots \\ t_N \end{bmatrix} \begin{bmatrix} m_1 \end{bmatrix} \approx \begin{bmatrix} d_1 \\ d_2 \\ \ldots \\ d_N \end{bmatrix}
$$

The observation equations are linear equations which have no solution, because the matrix on the left, here just the column vector of time values, is not square. Additional discussion of this point appears in Appendix A. Consequently, the observation equations must be converted to "normal" equations (there is only one in this simple example), which will have a solution. This conversion is done by multiplying the left- and right-hand sides by the transpose of the matrix on the left. After this step the approximately equals sign can be replaced by an equals sign because there will be as many equations as unknowns, with a single solution. This is the least squares estimate, denoted by $\hat{m}_1$.

$$
\begin{bmatrix} t_1 & t_2 & \cdots & t_N \end{bmatrix} \begin{bmatrix} t_1 \\ t_2 \\ \ldots \\ t_N \end{bmatrix} \begin{bmatrix} \hat{m}_1 \end{bmatrix} = \begin{bmatrix} t_1 & t_2 & \cdots & t_N \end{bmatrix} \begin{bmatrix} d_1 \\ d_2 \\ \ldots \\ d_N \end{bmatrix}
$$

Thus, after conversion of the above observation equations to a single normal equation, $\hat{m}_1$ has replaced m_1 and $=$ has replaced $\approx$. When the matrix multiplications in the above equation have been carried out we arrive at

$$
\left[\sum_{k=1}^{N} t_k^2 \right] \left[\hat{m}_1 \right] = \left[\sum_{k=1}^{N} t_k d_k \right]
$$

In fact the above equation just relates scalars, but we have added the large brackets to indicate that in general we have a matrix multiplying a column vector on the left-hand side and another column vector on the right-hand side. The normal equations are solved by finding the inverse of the matrix on the left and multiplying by it on both sides. With only one normal equation, the inverse is the reciprocal of a scalar, leading to

$$
\hat{m}_1 = \frac{\sum_{k=1}^{N} t_k d_k}{\sum_{k=1}^{N} t_k^2}
$$

which is identical to the maximum likelihood result, but requires no tedious derivatives. To show how easy it is to generalize to more parameters, remove the assumption that $m_0 = 0$, so that the observation equations are now

$$
\begin{bmatrix} 1 & t_1 \\ 1 & t_2 \\ \ldots & \ldots \\ 1 & t_N \end{bmatrix} \begin{bmatrix} m_0 \\ m_1 \end{bmatrix} \approx \begin{bmatrix} d_1 \\ d_2 \\ \ldots \\ d_N \end{bmatrix}
$$

The normal equations are obtained by multiplying the left- and right-hand sides by the matrix transpose to obtain

$$
\begin{bmatrix} N & \sum_{k=1}^{N} t_k \\ \sum_{k=1}^{N} t_k & \sum_{k=1}^{N} t_k^2 \end{bmatrix} \begin{bmatrix} \hat{m}_0 \\ \hat{m}_1 \end{bmatrix} \approx \begin{bmatrix} \sum_{k=1}^{N} d_k \\ \sum_{k=1}^{N} t_k d_k \end{bmatrix}
$$

The final result is obtained by finding the inverse of the symmetric matrix on the left and multiplying both sides by it. In summary, we seek a least squares estimate of a model (in the above case the model is the vector $m = [m_0, m_1]^T$, where the superscript T denotes the matrix transpose). A matrix G, in this case with two columns, is used in the observation equations to best predict d via the matrix equation $Gm \approx d$. The least squares solution is $\hat{m} = [G^T G]^{-1} G^T d$.

The transition from observation equations to normal equations involves multiplying by G^T. There is a geometrical interpretation of why this leads to the normal equations. Considering the case $Gm \approx d$ with three data and two unknowns, d and G have three rows, one for each data value, G has two columns, one for each of the two model parameters in the vector m. Let the two columns of G be called $g1$ and $g2$. In the vehicle speed problem just considered, $g1$ was all ones, and $g2$ was the set of time values. Here we generalize $g1$ and $g2$ to be any vectors, each with three components:

$$G = \begin{bmatrix} g1_1 & g2_1 \\ g1_2 & g2_2 \\ g1_3 & g2_3 \end{bmatrix} \tag{7.1}$$

Then the observation equations are given by

$$\begin{bmatrix} g1_1 & g2_1 \\ g1_2 & g2_2 \\ g1_3 & g2_3 \end{bmatrix} \begin{bmatrix} m_0 \\ m_1 \end{bmatrix} \approx \begin{bmatrix} d_1 \\ d_2 \\ d_3 \end{bmatrix}$$

The left-hand side can be rewritten as

$$m_0 \begin{bmatrix} g1_1 \\ g1_2 \\ g1_3 \end{bmatrix} + m_1 \begin{bmatrix} g2_1 \\ g2_2 \\ g2_3 \end{bmatrix}$$

The estimated model parameters $(\hat{m}_0, \hat{m}_1)$ must make the linear combination of $g1$ and $g2$ a best approximation to d. In linear algebra terms we seek to find an approximation to d within the vector space spanned by the columns of the matrix G, the column space of G. The approximation to d must therefore be a linear combination of $g1$ and $g2$, because every linear combination lies within a plane in three-dimensional space, in which $g1$ and $g2$ lie. This is illustrated in Figure 7.2. The best approximation to d makes the misfit vector $m_0 g1 + m_1 g2 - d$ shortest in length. The misfit vector will then be orthogonal to the plane defined by the two column vectors $g1, g2$. If the misfit vector is thus orthogonal, then the vector dot product of $g1$ with the misfit vector is zero, and the dot product of $g2$ with the misfit vector is also zero. These two conditions lead to the two normal equations. The two vector dot products are found by multiplying by the transposes of the two column vectors, so the normal equations for $\hat{m}$ are

$$g1^T [\hat{m}_0 g1 - d] = 0$$

and

$$g2^T [\hat{m}_1 g2 - d] = 0$$

which when combined into one matrix equation give $G^T [G\hat{m} - d] = 0$ or equivalently $G^t G\hat{m} = G^t d$. The least squares solution is obtained finally as

$$\hat{m} = (G^T G)^{-1} G^T d$$

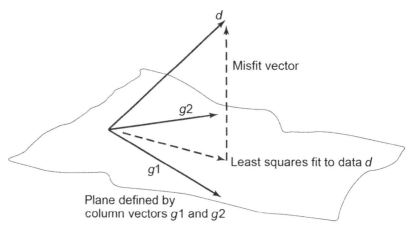

Figure 7.2 The two column vectors $g1$ and $g2$ of the G matrix define a plane, called the column space of the G matrix. The data vector d will not usually lie in that plane, so the data cannot be fitted exactly with the model. The least squares solution finds the coefficients of a linear combination of the column vectors $g1$ and $g2$ that best approximates d. This approximation will lie in the plane defined by $g1$ and $g2$ and will have a minimum distance to the arrowhead of the vector d. This geometrical justification for least squares does not depend on any assumption regarding the errors in the data vector d. In fact d may have no errors at all. In that case the difference between d and the linear combination of $g1$ and $g2$, $G\hat{m} - d$, is simply considered to be the model misfit.

With this geometrical view, we can understand how an under-determined problem might appear. Suppose all the time values are the same, so $t_1 = t_2 = t_3$. All the d values may still be different, because each is presumed to contain a different random error. The observation equations are then

$$\begin{bmatrix} 1 & t_1 \\ 1 & t_1 \\ 1 & t_1 \end{bmatrix} \begin{bmatrix} m_0 \\ m_1 \end{bmatrix} \approx \begin{bmatrix} d_1 \\ d_2 \\ d_3 \end{bmatrix}$$

The second column of the matrix is just t_1 times the first, so they are linearly dependent. The subspace spanned by the column vectors is now a line, not a plane. Forming normal equations by multiplying by the transpose

$$\begin{bmatrix} 3 & 3t_1 \\ 3t_1 & 3t_1^2 \end{bmatrix}$$

creates a matrix that does not have an inverse (its determinant is exactly zero), making it impossible to form the usual least squares solution. It is clear why this is so. In this case, neither the slope nor the intercept of a line passing through a point (corresponding to having data only at time t_1) is uniquely determined. There are infinitely many lines that satisfy the data, so, to obtain a solution, constraints are needed to regularize the problem. For example, if intercept m_0 is assumed zero, as was the initial assumption, we can solve for the speed, making the observation equations become

$$\begin{bmatrix} t_1 \\ t_1 \\ t_1 \end{bmatrix} \begin{bmatrix} m_1 \end{bmatrix} \approx \begin{bmatrix} d_1 \\ d_2 \\ d_3 \end{bmatrix}$$

and normal equations then produce the sensible solution

$$\hat{m}_1 = \frac{d_1 + d_2 + d_3}{3t_1}$$

where speed is thus determined using the average of the three d values. It is easy to spot the difficulty in this example, but with more complex problems there will be many data and many parameters. Some parameters in the model may be poorly determined or completely indeterminate. The tools of linear algebra, such as the singular value decomposition, can be used to examine the observation equations and to develop strategies to regularize the solution.

7.4 Weighted Least Squares

Now we show how to modify the least squares solution to account for variable data quality, so that poor (noisier) data have less importance in determining the solution. Information about data quality is contained in the square and symmetric data error covariance matrix C_d. At row k and column l, C_d contains the expected value of the product of data errors $n_k n_l$. On its diagonal are the variances of each n_k, $[\sigma_1^2, \sigma_2^2, \ldots, \sigma_N^2]$. Usually nothing is known about correlations among errors in the data, n_k, so C_d in most cases is taken to be a diagonal matrix, with zero off-diagonal elements. The weighting matrix is the inverse square root $C_d^{-1/2}$. In Appendix C this is identified as the matrix transforming correlated Gaussian random variables n_k into a new set of independent random variables. In the usual case, where C_d is assumed diagonal, $C_d^{-1/2}$ is also diagonal with reciprocal error standard deviations on each row, as follows:

$$\begin{bmatrix} 1/\sigma_1 & 0 & \cdots & 0 & 0 \\ 0 & 1/\sigma_2 & \cdots & 0 & 0 \\ 0 & 0 & 1/\sigma_3 & \cdots & 0 \\ \cdots & \cdots & \cdots & \cdots & \cdots \\ 0 & 0 & \cdots & \cdots & 1/\sigma_N \end{bmatrix}$$

Multiplying the observation equations by $C_d^{-1/2}$ on both the left- and the right-hand sides, each equation is scaled by the reciprocal standard deviation of the error, so each data value (and corresponding equation on the left-hand side) with a large error (large σ) is given a low weight. Multiplying by the transpose on the left, the weighted normal equations, $(G^T C_d^{-1} G)\hat{m} = (G^T C_d^{-1})d$, are solved in the usual way by multiplying both sides by the inverse of $G^T C_d^{-1} G$.

Weighted least squares requires information about the data error variances and covariances, but if this is unavailable then data editing is an alternative or supplementary strategy, in which suspect data are discarded. For example, if some values of d appear to be outliers, different from most other data, they might be removed, equivalent to using a weighting matrix in which their error standard deviation is taken to be infinity. Another strategy is data trimming, first computing a solution using all the data, then discarding those that appear to be in relatively poor agreement with the solution, a straight line in this case. The process may be repeated to successively trim the data. These strategies are useful when there are many data.

A simple application of weighted least squares is to show how to form a sensible average when the data are of unequal quality. Note that the observation equations for a simple average $\bar{d}$ of the values in data vector d (denoting the estimated average as $\hat{\bar{d}}$) are

$$\begin{bmatrix} 1 \\ 1 \\ \cdots \\ 1 \end{bmatrix} \begin{bmatrix} \bar{d} \end{bmatrix} \approx \begin{bmatrix} d_1 \\ d_2 \\ \cdots \\ d_N \end{bmatrix}$$

and multiplying on the left and right by the transpose results in

$$\hat{\bar{d}} = \frac{1}{N} \sum_{k=1}^{N} d_k$$

This confirms that a simple average is a least squares estimate. Now using $C_d^{-1/2}$ as the weighting matrix, assuming the usual case of uncorrelated errors but different standard deviations σ_k, the weighted observation equations are

$$\begin{bmatrix} 1/\sigma_1 \\ 1/\sigma_2 \\ \dots \\ 1/\sigma_N \end{bmatrix} \begin{bmatrix} \bar{d} \end{bmatrix} \approx \begin{bmatrix} d_1/\sigma_1 \\ d_2/\sigma_2 \\ \dots \\ d_N/\sigma_N \end{bmatrix}$$

Forming and solving the normal equations, we find that each data value is weighted by the reciprocal of its error variance. For example, for $N = 3$,

$$\hat{\bar{d}} = \frac{d_1/\sigma_1^2 + d_2/\sigma_2^2 + d_3/\sigma_3^2}{1/\sigma_1^2 + 1/\sigma_2^2 + 1/\sigma_3^2}$$

and the result generalizes for larger numbers of data.

7.5 Parameter Error Covariance Matrix

There is uncertainty associated with estimates $\hat{m}$, where

$$\hat{m} = (G^T C_d^{-1} G)^{-1} G^T C_d^{-1} d$$

due both to errors in the data and to insufficiency of the data available. An example of data insufficiency, above, was found in estimating both the slope and intercept of a line passing through a single point.

In standard least squares it is customary to use a parameter (model) error covariance matrix C_m to quantify the uncertainty due both to errors and data insufficiency. In this context, the model parameters are treated as Gaussian random variables. The diagonal elements of C_m are the variances of the model parameter errors, and the off-diagonal values are covariances of the errors among the various parameter estimates in the vector $\hat{m}$. Smaller variances indicate better-determined parameters, while large off-diagonal values indicate that the data have not been able to determine the individual parameters very well. Typically, if the off-diagonal terms are not too large, it is customary to report the square roots of the diagonal elements as standard deviations for each parameter. As described in Appendix C, a Gaussian random variable has a probability of about two-thirds of falling within plus or minus one standard deviation of its mean value. It is common (although not necessarily correct) to assume that estimated parameters behave as Gaussian random variables. Thus a stated confidence interval of two-thirds corresponds to plus and minus one standard deviation (from the parameter covariance matrix) surrounding the least squares estimate. This may provide a useful measure of confidence in simple problems, as in the example that follows, but in more complex problems with many parameters to estimate and diverse types of data, this approach will not be adequate. In more complex problems, Monte Carlo simulations with synthetic noisy data and known parameter values may be more fruitful in understanding the uncertainty in parameter estimates. Further discussion of confidence intervals appears in Appendix C.

When all the data are of the same type, with error variance σ^2, then $C_m = (G^T G)^{-1} \sigma^2$. To show that this is a sensible result, consider estimating the mean value of a set of N Gaussian random numbers in data vector d. The variance of each random number is σ^2 and all errors are independent. The least squares estimate is the simple average, as shown above.

To find C_m, note that $G^T G = N$, a 1×1 matrix, and its inverse multiplied by the data variance is σ^2/N. This confirms, as explained in Appendix C, that the variance of the average of N Gaussian independent random numbers is reduced by a factor N, so that the signal to noise variance ratio (SNR) improves by a factor N when multiple observations of the same quantity are averaged.

In the general case of a weighted least squares solution for model vector m, the data will be of different types, and there will be a data error covariance matrix C_d. Then the parameter covariance matrix associated with the estimate

$$\hat{m} = [G^T C_d^{-1} G]^{-1} G^T C_d^{-1} d = G^{-g} d$$

can be found in terms of the generalized inverse G^{-g}, where

$$G^{-g} = [G^T C_d^{-1} G]^{-1} G^T C_d^{-1}$$

The parameter covariance matrix is

$$C_m = [G^{-g}] C_d [G^{-g}]^T$$

which reduces to the previous result,

$$C_m = (G^T G)^{-1} \sigma^2$$

when C_d is diagonal with equal variances for all data.

7.6 Fitting Data to Sinusoids

A common least squares application is fitting time series or other data to standard functions such as sinusoids or polynomials. Coefficients are obtained by solving linear observation equations. Polynomial fitting was illustrated above. Fitting sinusoids is closely related to the DFT, and is common when analyzing climate data for average annual variations or sea level data for tidal variations. The DFT gives sinusoidal constituents when the frequency of interest is a Fourier frequency, the data are all of the same quality, and no data are missing. Least squares gives exactly the same result under these conditions but also works when the frequency of interest is not a Fourier frequency, when the data have variable quality or are missing, when the temporal sampling is not uniform, or when multiple data are taken at a single time.

To set up the observation equations, the frequency must always be given in the correct units. For example, if time is measured in days, then f must be in cycles per day, so, as an example, a regular annual variation will have frequency $f = 1/365$ cycles per day. The observation equations to fit a single-frequency sinusoid for time samples $t = [t_1, t_2, \ldots, t_N]$ are

$$\begin{bmatrix} \cos(2\pi f t_1) & \sin(2\pi f t_1) \\ \cos(2\pi f t_2) & \sin(2\pi f t_2) \\ \cdots & \cdots \\ \cos(2\pi f t_N) & \sin(2\pi f t_N) \end{bmatrix} \begin{bmatrix} c \\ s \end{bmatrix} \approx \begin{bmatrix} d_1 \\ d_2 \\ \cdots \\ d_N \end{bmatrix}$$

where c and s are the cosine and sine coefficients, respectively, and the observed data appear on the right-hand side. If there are two frequencies to be fitted simultaneously, say, both one and two cycles per year, two additional columns are added to the matrix on the left, leading to a 4×4 matrix in the normal equations. Ocean tide prediction is an example application, described in the next section.

A related problem is to find the frequency of a single sinusoid that best fits a time series. The additional parameter f appears in a non-linear way, unlike c and s, so the sum of squares $S(c, s, f)$ depends on three unknowns. However the problem is partly linear because, for any trial value of f, the coefficients c and s can be determined via the equations just presented, so $S(f)$ can be computed for trial values of f. The value $\hat{f}$ that makes $S(f)$ a minimum can easily be found in that case.

7.7 Ocean Tide Prediction

An important application of both the least squares and linear filter models is the prediction of tidal variations in water level at ports and harbors. Such predictions have historically been made using least squares fits of observed tidal variations to the known tidal forcing at specific frequencies, called the tidal constituents. A linear filter model is assumed for the ocean's response to tidal forcing, in which a single-frequency sinusoidal forcing produces a sinusoidal water level variation at exactly that frequency. This linear filter model is an approximation but has been in use for more than 150 years and has been proven to work reasonably well. More precise methods have come into use as computational power has increased.

The linear filter transfer function (known as the tidal admittance) at each constituent frequency is the ratio of the observed water level variations to the equilibrium tide height. The equilibrium tide height equals the tidal change in gravitational potential at that location divided by the local value of the gravitational acceleration. Ocean tides are far from being at equilibrium, so the tidal admittance is specified by both an amplitude and a phase. Once the admittance has been estimated from past obser-vations, it can be used to predict tides by adding the effects of all the tidal constituents at future times.

Ocean tides occur as Earth rotates within the gravity fields of the Sun and Moon. The tidal effect is proportional to the mass of the tide-causing body (Sun or Moon) and to the inverse cube of the distance to each. The lunar effect is more than double the solar effect. As Earth rotates it experiences approximately the same tidal force on the side facing the tide-causing body and on the opposite side. This produces two main tides per day, the lunar semi-diurnal (M2) and the solar semi-diurnal (S2). The period of the semi-diurnal tide M2 is not exactly 12 hours, but 12 hours plus an additional 25 minutes as the Moon advances in its orbit. The notation M2 and S2 was introduced in the early twentieth century by George Darwin, son of evolutionary biologist Charles Darwin. For these two constituents, the letters (M or S) identify the tide-causing body (Moon or Sun) and the integer 2 indicates approximately two high tides per day. Additional tidal constituents arise because Earth's rotation axis is inclined to the plane of the ecliptic (the plane of Earth's orbit), and because both Earth's orbit about the Sun and the Moon's orbit about Earth are eccentric. In addition, the lunar orbit is inclined to the plane of the ecliptic, and undergoes precession. Darwin's notation uses additional

Figure 7.3 Lord Kelvin's first tide prediction machine of 1872 adds together sinusoids of a number of discrete frequencies with adjustable amplitudes and phases. The machine adds the discrete frequencies of ten tidal constituents, to which the ten wheels correspond. Later versions of the machine added more wheels to include additional tidal constituents. Tide prediction machines were important in military and civilian applications, and remained in use until the 1960s. The addition of discrete-frequency sinusoids is similar to the addition of sampled discrete sinusoids performed by the inverse Discrete Fourier Transform in a digital computer. Tide prediction machines were programmed by adjusting the amplitude and phase of each constituent for future dates. The amplitude and phase responses (the tidal admittances) at a port or harbor were found by least squares fits of the constituent frequency sinusoids to actual tide-gauge records. The machine was operated by hand crank to turn the wheels and write the predicted tide on a paper chart. Photograph by William M. Connolley; source: en:Image:DSCN1739-thomson-tide-machine.jpg, licensed under the Creative Commons Attribution-Share Alike 3.0 Unported license.

letters for tidal constituents in the semi-diurnal band (integer 2) and the diurnal band (integer 1), and continues to be used for the largest constituents. The main constituents in the diurnal band are K1, O1, P1, Q1, and S1, with periods of 23.93, 25.82, 24.07, 26.87, and 24.00 hours, respectively. The main semi-diurnal constituents are M2, S2, N2, and K2, with periods of 12.42, 12.00, 12.66, and 11.97 hours, respectively.

   As outlined in the previous section, a least squares estimate of the tidal admittance is found by solving observation equations to fit sinusoids to a sampled time series of water level variations measured at a tide gauge, typically every hour. The columns of the observation equation matrix consist of a sine and cosine for each constituent theoretical equilibrium tide at a specified measurement time for that location. As a result, the dimension of the observation equation matrix in a typical case might be hundreds of rows by 20 or more columns. Prior to digital computers, least squares solutions of observation equations were accomplished with hand calculators, a highly laborious process. For observation equation matrices of large dimensions, as in the tidal prediction problem, this approach was not always practicable and so approximate methods were used. After the admittances have been calculated, a tide prediction is made by adding together all constituents at future times. Before digital computers, but continuing into the 1960s, mechanical tide prediction machines (analog computers) were used for this purpose. The most famous was developed by William Thomson (Lord Kelvin) in 1872 and is shown in Figure 7.3. Over the following decades many variants of the tide prediction machine were developed

Tidal prediction made substantial contributions to the success of military operations in World War II and to other military operations before and since. Among the greatest was prediction of tides for the beaches of Normandy, France, in early June 1944 prior to the Allied invasion of June 6, 1944 (D-Day). Various types of obstacles had been installed on the Normandy beaches to repel landing craft and located so as to be completely submerged at high tide. An invasion at low tide was essential to success. After solving for tidal admittances, predictions were made using a later model of Lord Kelvin's tide prediction machine that included more tidal constituents than the first model, shown in Figure 7.3. Given their strategic importance, tide prediction machine locations were a highly guarded military secret.

7.8 Seismic Tomography

Measurements of seismic wave travel times between sources and receivers are used to estimate the seismic velocity structure of the Earth at a variety of spatial scales. This application of least squares is called seismic tomography. Travel times are measured for waves whose travel paths (ray paths) through the Earth are known reasonably well. Because the ray path geometry is affected by seismic velocity, which bends rays according to Snell's law, in practice an initial velocity model is assumed and least squares is used to refine this model. At small spatial scales, cross-well tomography is used to monitor petroleum reservoir production or to estimate seismic velocities for geotechnical purposes. In volcanic settings, tomography can identify magma accumulations. At global scales, tomography has been effective in detecting subducted crustal slabs that have fallen deep into the mantle. Here we consider a simple example to show that seismic tomography can be posed as a linear least squares problem. This also illustrates the use of the parameter covariance matrix.

Seismic tomography typically adopts a discretized Earth model. The Earth is divided into volume elements (voxels), and the seismic velocity is estimated for individual voxels. The data consist of travel times for many ray paths each of which may pass through many voxels. An average seismic velocity model is used to calculate ray paths. Correlation filters are used to provide consistent measures of wave travel times, as will be discussed in Chapter 9. In practice, a data set may be incomplete (no ray paths passing through some voxels) but the example here does not deal with that case.

Consider a portion of the Earth divided into four cubic voxels, all with 1 km on a side, the unknown quantities being the four corresponding seismic velocities. The velocities $V1 - V4$ are to be estimated from six imperfect measurements of travel times. The locations of sources and receivers are known exactly, and the ray paths are taken to be straight lines from the midpoint of the outside of each voxel to the midpoint of the outside of another voxel. Referring to Figure 7.4, points 1 and 2 are surface locations and 3, 4, 5, 6 are well-bore locations. We denote the measured travel times as follows: $T36$, for example, is the travel time between point 3 and point 6. The travel time data have errors with known standard deviations of 0.1 s; the observed values in seconds are given by the vector

$$[T36 = 1.088, T45 = 0.917, T35 = 1.186, T46 = 0.977, T14 = 1.284, T25 = 1.499]$$

Travel time is distance divided by velocity, so the travel times do not depend linearly on velocity, but rather on its reciprocal. Reciprocal velocity is called slowness, and we will estimate the slowness s because it makes the problem linear. After estimating the slowness we can convert the values to velocity. We can set up the observation equations by recognizing that the travel time along each

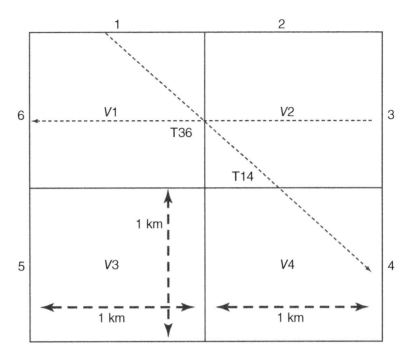

Four-voxel section of the Earth; travel times are measured between the midpoints of each voxel, each a 1 km cube. All measurement locations and ray paths are within the plane of the page. Example ray paths are show for travel times $T14$ and $T36$. All rays are assumed to be straight lines as shown by these two examples. The travel times are observed with errors that are known to be Gaussian with standard deviations of 0.1 seconds. The goal is to estimate $V1$, $V2$, $V3$, $V4$. The observation equations are linear in reciprocal velocity (slowness), so the slownesses are found by least squares and then inverted to express estimates in velocity.

ray path is the sum of the slowness multiplied by the distance traveled in each voxel. The simple geometry and assumed straight rays allow easy calculation of the distance associated with each travel time measurement. These distances appear (in units of km) on the rows of the observation equation matrix. Denoting the distance vector as $[s_1 = 1/V1, \ldots, s_4 = 1/V4]$, in the observation equations $Gm \approx d$ (m is the vector of slownesses, and d the data vector of the six travel times), the matrix G is formed as follows. In each of its rows are the distances in km traveled by the ray corresponding to the travel time in the column on the right. For example, in the first row, the ray for travel time $T36$ passes through voxels 1 and 2, traveling a distance of 1.0 km in each; these distances appear in columns 1 and 2. It does not pass through voxels 3 and 4, making the column 3 and 4 values equal to zero.

$$
\begin{bmatrix}
1.000 & 1.000 & 0 & 0 \\
0 & 0 & 1.000 & 1.000 \\
0 & 1.118 & 1.118 & 0 \\
1.118 & 0 & 0 & 1.118 \\
0.707 & 0.707 & 0 & 0.707 \\
0.707 & 0.707 & 0.707 & 0
\end{bmatrix}
\begin{bmatrix}
s_1 \\
s_2 \\
s_3 \\
s_4
\end{bmatrix}
\approx
\begin{bmatrix}
T36 \\
T45 \\
T35 \\
T46 \\
T14 \\
T25
\end{bmatrix}
$$

The least squares solution yields the following slowness values in units of s/km:

$$\begin{bmatrix} s_1 \\ s_2 \\ s_3 \\ s_4 \end{bmatrix} = \begin{bmatrix} 0.6738 \\ 0.5562 \\ 0.6122 \\ 0.3081 \end{bmatrix}$$

Taking their reciprocals provides velocity estimates in km/s:

$$\begin{bmatrix} V1 \\ V2 \\ V3 \\ V4 \end{bmatrix} = \begin{bmatrix} 1.4841 \\ 1.7981 \\ 1.6334 \\ 3.2455 \end{bmatrix}$$

Because each of the six travel time measurements involves rays passing through several voxels, it is clear that we cannot uniquely determine the velocity in each. We need to describe the uncertainty associated with this limitation (determined by the way in which the data set samples the Earth), and also the uncertainty associated with travel time errors. We compute the parameter covariance matrix, as described in Section 7.4 to do this. Working with slowness as the estimated parameter, the covariance matrix C_m is found to be, in units of $0.01 \times (\text{s/km})^2$,

$$\begin{bmatrix} 1.53 & -1.27 & 0.92 & -1.08 \\ -1.27 & 1.53 & -1.08 & 0.92 \\ 0.92 & -1.08 & 1.18 & -0.82 \\ -1.08 & 0.92 & -0.82 & 1.18 \end{bmatrix}$$

The diagonal elements of C_m tend to be larger than the off-diagonal, but not in all cases. Nevertheless, it is customary to take the diagonal values as estimates of the variances of the errors in the slowness estimates $[s_1, \dots, s_4]$ on rows 1 to 4. Assuming this, we can estimate an associated plus or minus one standard deviation (with approximate 67 percent confidence) for the slowness and take the reciprocals of the upper and lower bounds to find confidence intervals for velocity that reflect both the errors in travel time measurement and the limitations of the available data set.

7.9 A Model for Global Sea Level Change

The global mean sea level change has been routinely measured by satellite radar altimeters since the Topex mission was launched in 1992 and placed in orbit at an altitude slightly above 1300 km. Several generations of satellites have been placed into service over the decades since then, at similar altitudes. Measurement involves a downward pointing radar which illuminates the sea surface over a footprint on the order of 10 km diameter at the sea surface. The round-trip travel time of the radar pulse is converted to distance by multiplying by the speed of light, with corrections for conditions that affect it, such as water vapor in the troposphere and free electron density in the ionosphere. The measurements are validated and calibrated using tide gauges and other methods. Altimetry satellites must be precisely located relative to Earth's center. Starting with

Topex, precise satellite orbit determination methods have made use of the Global Positioning System (GPS), as discussed in Chapter 9. Other global navigation satellite systems (GNSSs) and laser ranging from ground stations have been used subsequently to provide precise orbits. Most altimetry satellites, starting with Topex, were placed in orbits that covered most of the global oceans but excluded very high latitudes. High latitudes were omitted, in part, because the altimetry measurement becomes more difficult in the presence of floating ice and nearby land. Orbits for many successor missions have been chosen to provide measurements every 10 days over the same ground tracks.

The global mean sea level time series in Figure 1.1 was combined from multiple satellite missions, starting with Topex-Poseidon in 1992, with a sample interval of 10 days. Each sample is an average of many millions of individual range measurements over the oceans, so the actual measurement error associated with a single sample must be negligible, although other types of errors which are not random may be present. Figure 1.1 and similar altimetry time series are considered to be both a key measure of contemporary climate variability and an essential measurement for assessing impacts on human activities in coastal regions. It is sensible to develop a time series model to account for variations in Figure 1.1. Such a model may improve understanding of important physical processes in the climate system, and provide a basis for predictions of future sea level change. We can use least squares to develop a model, using physical reasoning, intuition, and inspection of residuals to guide development. By adopting least squares we seek sensible estimates of model parameters, but recognize that deviations from a proposed model may represent model misfit and are not necessarily due to data errors. Model misfit may include other signals which are not yet understood. This suggests that a maximum likelihood motivation for least squares may not apply in this case because Gaussian random errors are not likely to be the main cause of departures from a straight line or other model elements such as seasonal frequency sinusoids.

Figure 1.1 suggests that the dominant change in global sea level is a steady rise over time. We can estimate the rate of global sea level rise by fitting a polynomial to the time series. If we fit a linear (degree-1) polynomial, we obtain an average rate of sea level rise in mm/year. If the model were extended to a quadratic polynomial (degree-2), we would allow for the possibility of acceleration or deceleration in the rate. We adopt a degree-1 model here. The question of whether there is significant acceleration in sea level rise is a continuing research topic. After estimating the rate of sea level rise, we subtract the linear trend from the original time series, leaving a detrended residual (shown in Figures 1.1 and 7.5) for further analysis and model development.

The detrended residual sea level time series in Figure 1.1 shows seasonal oscillations. These are an expected consequence of several processes, including seasonal variations in the amount of water in the oceans due to the terrestrial water cycle, which exchanges water between land and oceans, and seasonal ocean temperature variations, which affect thermal expansion. Therefore, a second element of the model will be to fit sinusoids at seasonal frequencies. Finally, we will examine the non-seasonal residual variations in an effort to understand what may be causing them.

Figure 7.5 shows residual sea level time series obtained in the process of model development. The top curve is the detrended time series. The slope of the linear trend is variable depending on whether seasonal and other terms are first removed before computing the slope. This is left as an exercise. In published studies, a global sea level rise rate is often assigned a confidence interval. A typical 90 percent confidence interval is plus or minus 0.4 mm per year. This is greater than the variability associated with differences that arise depending on whether we estimate the slope before or after

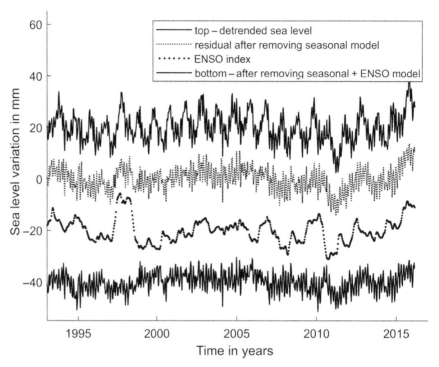

Figure 7.5 Global sea level time series from satellite radar altimetry; the curves are offset for clarity. Figure 1.1 shows both the original time series and the detrended version. The top curve here also shows the detrended version. Second from the top is the residual after the seasonal model is removed from the detrended time series. The seasonal model removed includes sinusoids of frequencies up to 6 cycles per year. The third curve shows the ENSO index from Figure 1.4. It appears to be correlated with the non-seasonal residual, suggesting that the model be adjusted to include a variation proportional to the ENSO time series. Addition of one more model parameter proportional to the ENSO index leaves the residual time series at the bottom. The seasonal model alone explains about 39 percent of the variance of the top curve, while addition of the ENSO time series to the model increases the explained variance to 56 percent. This large increase in explained variance with the addition of one more parameter confirms that including the ENSO index was a useful addition to the model.

removing seasonal or other signals. The manner in which a confidence interval is obtained is the subject of current research; it should take into consideration various sources including both random error and systematic errors in the measuring system. Interpreting a confidence interval is difficult, partly because we are uncertain about the correctness of the polynomial model (linear, quadratic, or other).

The second curve from the top in Figure 7.5 shows the residual after removing a seasonal model with frequencies up to 6 cpy (cycles per year). The sinusoidal coefficients in the following array are for the frequencies indicated in the first column in the table below, with cosine and sine coefficients in the second and third, respectively, in units of mm. The phases of the cosine and sine refer to the time of the first Topex data point, at 1992.9595. Diminishing amplitudes at higher frequencies confirm that seasonal variations are dominantly at 1 and 2 cpy:

cpy	cosine	sine
1	1.463	−4.793
2	−1.164	−0.429
3	−0.365	−0.088
4	0.230	0.045
5	−0.164	−0.207
6	0.057	−0.173

Figure 7.5 shows that the ENSO index, third from the top (and appearing earlier in Figure 1.4) looks correlated with the seasonal residual. A connection with ENSO would not be suprising, so we can revise the model to add one more parameter to allow a contribution from the ENSO index. This adds one more column to the observation equation matrix. Now the matrix contains the 12 columns of cosines and sines at the sample times, plus the ENSO time series. The residual is the bottom curve in Figure 7.5.

We might ask whether the addition of an ENSO index term was a good idea. In the process of model development one often addresses this question by assessing how much more variance is accounted for by including an additional parameter, in this case a term proportional to the ENSO index. We can calculate the percentage of explained variance (ev) for any model. If the original time series variance is σ^2 and the variance of the residual after fitting the model is σ_r^2 then the percentage of explained variance is

$$ev = 100 \times \frac{\sigma^2 - \sigma_r^2}{\sigma^2}$$

For the seasonal model ev is about 39 percent, while addition of the ENSO index to the model increases ev to about 56 percent, a large improvement that justifies its inclusion in the model.

7.10 Chapter Summary

Least squares is among the most important data processing and analysis tools in geophysics and other disciplines. The main points are as follows.

- The least square fitting of a time series to functions such as polynomials is a maximum likelihood (most probable) estimate when the misfit between function and time series is due to Gaussian (normally) distributed random errors.
- When viewed as a solution to over-determined simultaneous linear observation equations, least squares yields a data misfit vector orthogonal to the column space of the linear equation matrix.
- Least squares and the DFT provide identical results for a uniformly sampled time series when a Fourier frequency sinusoid is being fitted.
- Least squares is useful for fitting sinusoids in problems that the DFT cannot address: fitting other than Fourier frequency sinusoids; fitting to time series that are not uniformly sampled in time or have missing data; and for cases where time series values are of variable quality.
- Variable quality data are described by a data error covariance matrix. A weighting matrix is used to multiply the observation equation matrix and data vector, to give the better data more importance in the solution. Usually the weighting matrix is the inverse square root of the data error covariance

matrix. Formally, this weighting matrix is tied to an assumption of Gaussian errors, but even if that assumption is not true, it provides sensible results.

- Errors in least squares parameter estimates may be due both to insufficient data and to errors in the data. If the estimated parameters are treated as Gaussian random variables, then a useful description of errors is the model parameter covariance matrix.

Exercises

7.1   **Ordinary and Weighted Least Squares.** Assume you have three measurements of gravity $[g_1, g_2, g_3]$ taken with a gravity meter at the same location on different days. The error of each may differ, depending on the wind conditions at the time of measurement. Suppose that the standard deviations of the errors in the three readings are $[\sigma_1, \sigma_2, \sigma_3]$ and that the errors are independent. The data error covariance matrix has on its diagonal the square of these standard deviations and zero off the diagonal, since the errors are assumed independent.

A. Write down the unweighted observation equations and find their solution (by hand), to show that the least squares solution for the weighting of the three data values is 1/3, so the data are combined by simply averaging the three measurements.

B. Write down the weighted observations equations multiplying those in part A on both sides by the inverse square root error covariance matrix, taking it to be diagonal. Solve by hand to show that the weights are reciprocal variances, normalized to add to 1.

C. Let the standard deviations of the three gravity readings be [2, 4, 10] (microgals), so that g_1 is better than the other two readings and should be more important in the weighted average, and g_3 should be given the least weight. Find the weights for the three measurements $w = [w_1, w_2, w_3]$ such that the weighted average of $[w_1 g_1 + w_2 g_2 + w_3 g_3]$ is an estimate with minimized noise variance, and such that the weights add to 1.

D. Find the error reduction in decibels achieved with the weighted average in part C, compared to a simple average obtained using the error standard deviations given in C.

7.2   **Seismic Tomography.**

A. Repeat the calculation of the parameter error covariance matrix in the seismic tomography example illustrated by Figure 7.4 to find the approximate confidence intervals for velocity. The upper confidence interval for each slowness is taken as the reciprocal of the estimated slowness for that voxel plus the standard deviation of the slowness, and the lower confidence interval is the estimate minus the slowness standard deviation. The velocity confidence intervals are found from the reciprocal of these slowness values.

B. The parameter error covariance matrix given in the text is proportional to the stated travel time error variance of $0.01 \, \text{s}^2$. Suppose however that the travel time error variance is unknown. This is a common situation, and in this case the variance of the data errors must be estimated. The usual way is to first find the misfit vector $G\hat{m} - d$. In this case it is the set of the six residual times. An estimate of error variance is found by computing the sum of the six squared residual values. This is the sum of squares, and so is converted to variance by dividing in this case by 2, not by 5 or 6 as in the usual variance calculation. The divisor 2 equals the number of data (6) minus the number of parameters fit in the model (the four slownesses). You can see that this is sensible, because if we only had four travel time measurements, we could fit them exactly with four slownesses, and the residuals would all be zero. So, the divisor should be the number of data residuals minus the number of parameters being fitted. Of course this will give a different answer, but nevertheless use this method to estimate the travel time error variance from the residuals and compare it with the stated value of 0.01 s^2.

7.3   **Sea Level Model.** Use the 10-day samples of global sea level change from satellite altimetry, and the ENSO index time series for the same period, to construct a model for global sea level variations using least squares as described in the text.

A. Show that slightly different sea level trend rates are obtained depending on whether the linear rate is first determined, and then subtracted before fitting other elements of the model such as seasonal sinusoids, or, alternatively, if the linear rate is included in a single model along with the sinusoids and the ENSO index. In case of a single model, the observation equation matrix would have 15 columns, two for the linear polynomial trend (slope plus intercept), 12 for the seasonal model (six frequencies with corresponding cosine and sine coefficients), and one column for the ENSO index.

B. Plot a histogram of the sea level residual time series values to confirm its general appearance as a bell-shaped curve. Histograms of residuals are often plotted to test goodness of the model fit, with an underlying assumption that Gaussian random errors are the cause of model misfit. For Gaussian random errors, a bell-shaped histogram would be expected. In this present case the measurement errors are unknown and the model misfits, that is, the signals not explained by the model, are treated as errors. Their histogram suggests a bell shape, but it would not by itself inform you that the residual time series (the bottom curve in Figure 7.5) is not due to random error. Instead, inspection of the autocorrelation of the residual time series confirms this. So, find the autocorrelation of the residual time series and plot it for lags of up to 100 samples to show that it differs from that for white noise.

C. Compute an error standard deviation for the estimated slope by using the variance of the misfit vector (in units of mm) to scale the parameter error covariance matrix when only a slope and intercept are being fitted to the sea level time series, and for other cases when the slope is estimated from the data after seasonal signals and/or ENSO are removed. Plus and minus two standard deviations would be quoted as a nominal 90 percent confidence interval for the rate of sea level rise. This formal error estimated from the data alone is smaller than the published estimates since it does not account for measurement error and other sorts of errors, which are the subject of continuing research.

8 Linear Filter Design

Linear time domain filters are useful in data processing and as models of physical systems. This chapter reviews examples of digital filter design methods, with applications to data processing, as approximations to physical systems (a seismometer is used as an example) and as a description of physical processes such as echoes and reverberations, gravity anomalies, and ground motion amplification in an earthquake. The focus here is on time domain filters. However, many of these problems may also be addressed in the frequency domain using the Discrete Fourier Transform.

8.1 Introducing the Z Plane

Given the general form of an ARMA filter (Section 5.1),

$$a_0 y_t = b_0 x_t + b_1 x_{t-1} + \cdots + b_n x_{t-n} - a_1 y_{t-1} - \cdots - a_m y_{t-m}$$

the design problem is to set the filter order $[m, n]$ and find numerical values for $a_t = [a_0, a_1, \ldots]$ and $b_t = [b_0, b_1, \ldots]$. For a given design objective, usually specified by the transfer function (Section 5.5), there may be many different choices for these coefficients, so digital filter design is often as much an art as a science.

If the goal is to develop a filter with specific transfer function properties, recalling Sections 4.6 and 5.5, we first recognize that Z transform polynomials of MA and AR coefficient arrays provide a compact expression for a digital filter transfer function

$$L(Z) = L(f) = \frac{B(Z)}{A(Z)}$$

where $Z = \exp(-2\pi i f)$ and f ranges over the Nyquist band $[-1/2, 1/2]$. Suppose now that Z is allowed to be any complex number. Then we can define a complex Z plane, with its origin at the center of the unit circle. Every point on the unit circle corresponds to a real value of frequency f in the Nyquist band. For example, $f = 0$ corresponds to $Z = 1+0i$, while $f = 1/4$ (half the Nyquist) is at $Z = 0 - i$. (Note that the alternate definition $Z = \exp(+2\pi i f)$ is used by some authors.) We can then say that $L(f)$ is an evaluation of $B(Z)/A(Z)$ on the unit circle. Figure 8.1 shows how the Nyquist frequency band occupies the circumference of the unit circle. Additional elements of Figure 8.1 are discussed in the next section.

If we factor both the $B(Z)$ and $A(Z)$ polynomials, then the transfer function is

$$L(Z) = \frac{C(Z - q_1)(Z - q_2) \cdots}{(Z - p_1)(Z - p_2) \cdots}$$

where C is a scalar, $q_1, q_2, \ldots$ are the zeros of the transfer functions, and $p_1, p_2, \ldots$ are its poles. A pure MA filter is called all-zero, providing yet another name for MA filters, in addition to finite impulse response (FIR). An infinite impulse response (IIR) AR filter may be called all-pole and an

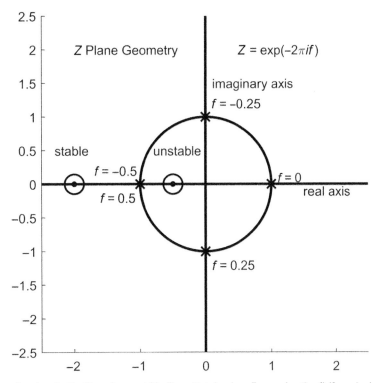

Figure 8.1   The Z plane is a complex plane for the Z transform variable. The unit circle where $Z = \exp(-i2\pi f)$ (for real values of frequency f) plays a critical role in understanding Z plane geometry. The linear filter transfer functions are ratios of Z transform polynomials. Z plane locations of roots of these polynomials relative to the unit circle provide a geometrical explanation for transfer function behavior. The roots of the denominator polynomials $A(Z)$, associated with the AR coefficients, are called poles. The roots of the numerator polynomials $B(Z)$, associated with the MA coefficients, are called zeros. Unit circle locations corresponding to a few values of frequency ($f = 0, \pm0.25$, and ±0.5) are labeled with crosses. Poles amplify the frequencies associated with nearby portions of the unit circle. Zeros attenuate the frequencies associated with nearby unit circle locations. If a pole is located on the interior of the unit circle, then the filter is unstable. When $A(Z)$ is factored into a product of poles, one or more may be located inside the unit circle, but even a single pole inside the unit circle will render the entire filter unstable. Zero locations indicated by the symbol $\odot$ are shown for the two MA filters $y_t = 2x_t + x_{t-1}$ and $y_t = x_t + 2x_{t-1}$. These two filters have exactly the same power transfer function but different phase behavior, as shown in Figure 8.2. In general every single pole or single zero filter will have a filter twin, with the same power transfer function and pole or zero at the polar reciprocal position (radius from the origin is reciprocal, located along the same radial line). The zero location for $y_t = x_t + 2x_{t-1}$ is inside the unit circle, so its inverse (which would have a pole at this location) would be unstable, meaning the filter is not invertible. The MA filter itself is stable but has maximum-phase, as shown in Figure 8.2. It is possible to convert an unstable digital filter (with pole inside the unit circle) into a stable one by replacing the pole with its polar reciprocal value.

ARMA filter is mixed-pole-zero. In the usual case of real-valued filter coefficients, $A(Z)$ and $B(Z)$ are polynomials with real coefficients. A well-known result (related to the fundamental theorem of algebra) is that roots of polynomials with real coefficients are either real-valued or occur as complex conjugate pairs. Poles and zeros will therefore be arranged symmetrically in the Z plane above and below the real axis (as conjugates), or else will lie on the real axis. The pole and zero symmetry about the real axis in the Z plane is associated with the Hermitian symmetry of transfer functions, providing equal transfer function gain to positive frequencies on the lower half of the unit circle and to negative frequencies on the upper half. In a qualitative sense we note that the transfer function will have reduced magnitude for frequencies (points on the unit circle) near a transfer function zero, and greater magnitude for frequencies near a pole. A geometrical understanding of the relationship between transfer function gain and pole and zero proximity to the unit circle provides a useful and intuitive approach to filter design. This is illustrated by examples later in this chapter.

8.2 *Z* Plane Geometry – Stability and Invertibility

The transfer function of a cascade of filters is the product of their transfer functions, so, when $A(Z)$ and $B(Z)$ are factored, the product of factors is recognized as a cascade of single-zero (MA) and single-pole (AR) filter stages. There are important consequences for the stability of a digital filter and the existence of a stable inverse, depending on whether the transfer function poles and zeros are located inside or outside the unit circle. This can be illustrated using a pair of single-zero (MA) filters and their single-pole inverses as examples. Figure 8.1 shows the Z plane locations of the poles and zeros for two filters and their inverses. Filter 1 is given by

$$y_t = 2x_t + x_{t-1} = (2, 1) * x_t$$

$$L(Z) = 2 + Z$$

with a zero at $Z = -2$ whose inverse is the AR filter

$$y_t = \frac{1}{2}x_t - \frac{1}{2}y_{t-1}$$

$$L(Z) = \frac{1}{2 + Z}$$

also with a pole at $Z = -2$. The impulse response of the inverse filter is $[\frac{1}{2}, -\frac{1}{4}, \frac{1}{8}, \ldots]$, showing that it is stable.

Filter 2 is given by

$$y_t = x_t + 2x_{t-1} = (1, 2) * x_t$$

$$L(Z) = 1 + 2Z$$

with a zero at $Z = -\frac{1}{2}$ and inverse filter

$$y_t = x_t - 2y_{t-1}$$

which has transfer function $L(Z) = 1/(1 + 2Z)$ with a pole at $Z = -\frac{1}{2}$. The impulse response of its inverse filter is $[1, -2, 4, -8, \ldots]$, showing it to be unstable.

As an aside, there is an interesting consequence that follows from expressing transfer functions as Z polynomials. We can use polynomial long division to convert this inverse filter transfer function (and also the previous stable one) into an infinite-length Z polynomial. It is not hard to show that

$$L(Z) = \frac{1}{1 + 2Z} = 1 - 2Z + 4Z^2 - 8Z^3 + \cdots$$

The right-hand side is just the Z transform of the impulse response time series, in this case a series that diverges to infinite values because the filter is unstable.

The moving average Filters 1 and 2 involve convolution with the MA coefficients; these are described as a length-2 time series, known as a couplet. The first couplet $(2, 1)$ with largest value first is called front-loaded, minimum-delay, or minimum-phase. The second couplet $(1, 2)$ is back-loaded, maximum delay, or maximum-phase. The inverse filter for the minimum-phase MA filter is stable, while the inverse of the maximum-phase MA filter is unstable. Figure 8.1 shows the geometry of the two MA filters in the Z plane. It is conventional to use the symbols $\odot$ for zeros and $\otimes$ for poles. The zeros for the two filters are at reciprocal distances from the origin. If these zeros were complex-valued, it would be clear that they are polar reciprocals of one another, at reciprocal distances, lying along the same radial line from the origin.

The power transfer functions of the two MA filters are found by multiplying each transfer function by its complex conjugate. Here $Z = \exp(-2\pi i f)$, with conjugate (denoted by a superscript $*$) equal to $Z^* = \exp(2\pi i f) = Z^{-1}$, making the product $ZZ^* = 1$. For the minimum-phase filter the power transfer function is

$$(Z + 2)(Z^* + 2) = (1 + 2(Z + Z^*) + 4) = 5 + 4\cos(2\pi f)$$

when $Z = \exp(-2\pi i f)$ and the positive and negative powers of Z are to form the cosine. For the maximum-delay filter the power transfer function is found to be exactly the same:

$$(2Z + 1)(2Z^* + 1) = (4 + 2(Z + Z^*) + 1) = 5 + 4\cos(2\pi f)$$

Figure 8.2 shows the power and phase transfer functions of Filters 1 and 2. Being MA filters, they amplify low frequencies relative to high because their zeros are located on the unit circle near $Z = -1$ where the Nyquist frequency is located; the lower frequencies, near $Z = 1$, are farther from the zeros, so are less attenuated. The opposite is true for the inverse filters, for which the frequencies closer to poles are amplified.

This example shows that the two different filters have exactly the same power transfer function. Either filter twin might be chosen but to guarantee stability, or the existence of a stable inverse, the minimum-phase twin is chosen. This conclusion extends to a general ARMA filter, with transfer function polynomials $A(Z)$ and $B(Z)$, which can be represented as a cascade of single-zero and single-pole filters. For a stable filter and/or inverse, every factor (that is, every filter stage) must be minimum-phase. Therefore, all poles and zeros must lie outside the unit circle. If even one pole of $A(Z)$ lies inside the unit circle it will have an unstable impulse response, because when its impulse response is convolved with all others in the cascade, the result will grow in an unbounded way. (With the alternative definition $Z = \exp(+i2\pi f)$, stable or invertible filters have poles in the interior of the unit circle.)

Some day, you might be asked to find the inverse of an MA filter which contains one or more filter stages that are not minimum-phase. No stable inverse exists in this case. However, you can find a pretty good inverse filter that is stable by substituting each zero of the MA filter that lies inside the unit circle with its twin outside the unit circle. This requires factoring the transfer function Z polynomial, but computer algorithms are available for that task. The twin will be at the polar reciprocal location, and the power transfer function of the modified MA filter will remain unchanged but, with all zeros outside the unit circle, its AR inverse will be stable. In other situations you might be given a power transfer function, power spectrum, or autocorrelation, and asked to find the minimum-phase time function to which it corresponds. Besides polynomial factoring, other techniques exist to address this spectral factorization problem. One solution is to to fit AR filter models using the Yule–Walker equations, to be described in Chapter 9. The autoregressive filters (prediction error filters) from the Yule–Walker equations are always minimum-phase.

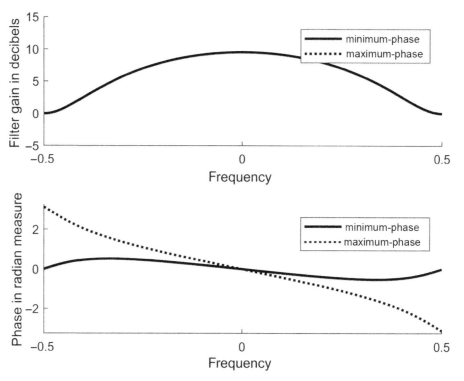

Figure 8.2   Power and phase transfer functions for the two MA filters, one minimum-phase (zeros outside the unit circle), the other maximum-phase (zero inside the unit circle). The power transfer functions are the same. Minimum-phase filters show no phase discontinuity at the Nyquist frequency; only the minimum-phase filter has a stable inverse. Some authors use a different definition, choosing to let $Z = \exp(+i2\pi f)$. In this case the stable region of the Z plane is inside the unit circle.

8.3   Notch Filter Design Using Z Plane Geometry

The use of Z plane geometry methods involves placing poles near the unit circle locations of the frequencies to be amplified, and zeros near those to be attenuated. A simple example is the development of a filter to remove electric utility line noise at 60 Hz from recorded data. The goal is thus to develop a "notch" filter whose transfer function will be zero at 60 Hz and approximately unity at other frequencies. Rather than using a notch filter, other approaches to this problem might include using least squares to fit and subtract a 60 Hz sinusoid or filtering with the DFT if 60 Hz is a Fourier frequency. Both methods would require that the data be recorded.

Let the digital sample interval be 4 milliseconds, so that the Nyquist frequency is 125 Hz, and frequencies of $\pm 60\ Hz$ correspond to points on the unit circle at $Z = \exp(-\pi i (60/125))$, for +60 Hz, and $Z = \exp(+\pi i (60/125))$ for −60 Hz. If we place zeros at these locations, the transfer function will be exactly zero at 60 Hz. Placing zeros on the unit circle, rather than outside it, means that the filter will not be invertible, because it completely removes information about 60 Hz. Invertibility is not an important consideration here, however. To simplify the notation, let $Z_{60} = \exp(-\pi i (60/125))$ and $Z_{-60} = \exp(+\pi i (60/125))$, where Z_{60} is the point on the unit circle corresponding to 60 Hz and Z_{-60} is the point corresponding to −60 Hz. The transfer function is

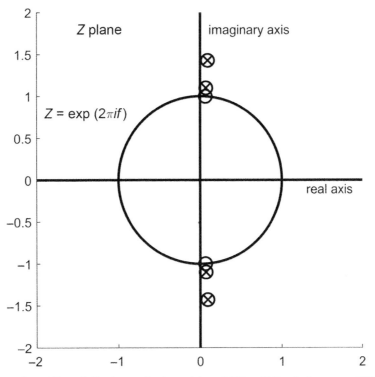

Figure 8.3 The Z plane geometry for a 60 Hz notch filter for sampling interval $\Delta t = 0.004$ s, with Nyquist frequency ± 125 Hz located at $Z = -1$. A single zero at positive and negative 60 Hz on the unit circle removes those frequencies, but by itself is a poor notch filter because it attenuates nearby frequencies as well. Placing a pole just outside the unit circle at ± 60 Hz creates a sharper notch. The poles at radii 1.1 and 1.43 are both shown, and their corresponding transfer functions, shown in Figure 8.4, confirm that a radius of about 1.1 is a good choice.

$$(Z - Z_{60})(Z - Z_{-60})$$

which when simplified is $Z^2 - (0.1256)Z + 1$, so the filter equation is

$$y_t = x_t - (0.1256)x_{t-1} + x_{t-2}$$

Although the transfer function is indeed zero at 60 Hz, it also reduces amplitudes at nearby frequencies. It is therefore a poor notch filter, but it can be improved by the addition of a pole just outside the unit circle near each zero. Figure 8.3 shows the Z plane geometry. The influence of the pole and of the zero will tend to cancel, except at frequencies (unit circle locations) near ± 60 Hz. We can control the width of the notch by adjusting the proximity of the pole to the unit circle; making it closer will create a sharper notch. Figure 8.4 shows three power transfer functions, one with no poles, a second with pole radius of 1.1, for which the transfer function is

$$\frac{(Z - Z_{60})(Z - Z_{-60})}{(Z - 1.1Z_{60})(Z - 1.1Z_{-60})}$$

and a third with a pole radius of 1.43, having a similar transfer function, with 1.43 replacing the factor 1.1.

A pole radius of 1.1 creates an excellent notch at 60 Hz, with the filter gain approximately constant at most other frequencies. A drawback is that this filter has a longer impulse response than the

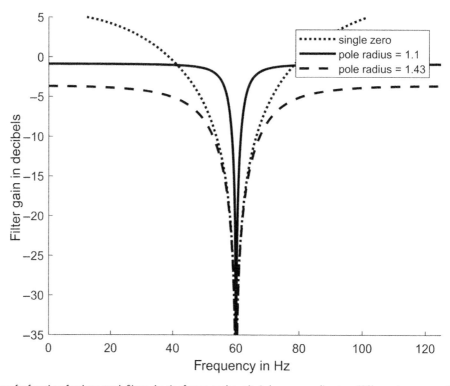

Figure 8.4 Power transfer functions for three notch filters. A pair of zeros on the unit circle corresponding to ±60 Hz creates a poor notch filter. The transfer functions for poles at radii 1.1 or 1.43 show that the notch can be sharpened by bringing it closer to the unit circle. The resulting two-zero and two-pole filter may be implemented as a zero-phase filter by passing the filter across the time series in both the forward and reverse directions, as described in the text.

other two. This is generally not desirable because if noise is present, there will be longer-duration transient effects in the filtered data. This sort of situation is generally the case in digital filter design. Typically, one must trade off various objectives, such as matching a desired transfer function, keeping the impulse response short, and reducing computational effort by keeping the filter order to a minimum.

A final consideration is phase response. We want the notch filter to be zero-phase, because the goal is to reject 60 Hz but to retain other frequencies unchanged, both in amplitude and phase. While the phase was not considered in placing the poles and zeros, any digital filter can be implemented as zero-phase by applying it twice in a cascade. The first application filters the time series in the usual way. Then this filtered series is reversed in time, and the filter is applied again. Finally this result is reversed so that time proceeds in the normal direction. The two-pass (forward–reverse) zero-phase filter power transfer function is the squared modulus of the single-pass filter, and the effect of filtering twice is to make the notch even sharper.

8.4 Differential Equation to Digital Filter Equation

Here we develop a digital filter that behaves like a damped harmonic oscillator, with decaying sinusoidal impulse response. This is a simplified description of a vertical seismometer. The filter can be used to simulate the effects of the seismometer on the data, and the inverse filter can be used to remove them from the observed seismic waves. The process involves identifying poles in the complex plane associated with continuous frequency and mapping them into Z plane poles to find the digital filter coefficients. Manufacturers typically provide information about the poles and zeros of their seismometers, which incorporate complex feedback circuits to make them sensitive to a wider range of frequencies than the simple mass-on-spring instrument described here.

The starting point is the linear differential equation describing a damped mass-on-spring seismometer. The general approach is to identify poles and zeros in the complex-frequency plane determined by the differential equation, which are to be mapped to the Z plane. This approach is applicable to any system described by a linear differential equation.

Let Earth's center of mass be the origin of an approximately inertial reference frame, in which the vertical coordinate z is used to measure the position of a mass m. In the same reference frame, let x be the vertical position on Earth's surface at which the seismometer is mounted, so that the displacement of the mass relative to Earth's surface is $z - x$. The damping of the mass's motion is proportional to its vertical velocity relative to Earth's surface. The force of the spring is proportional to the product of the spring constant k and the displacement of the mass relative to Earth's surface. The equation of motion for the mass in its inertial frame requires that the forces on the mass must add to zero. The following equation gives the sum, from left to right, of the inertial, damping, and spring forces, set equal to zero.

$$m\frac{d^2z}{dt^2} + (damping)\frac{d(z-x)}{dt} + k(z-x) = 0$$

where *damping* is a place-holder for a scalar coefficient to be inserted below.

Let $y = z - x$ be the measured displacement of the mass relative to the Earth's surface; then, rewriting the equation in terms of y, we have

$$\frac{d^2y}{dt^2} + (\omega_0/Q)\frac{dy}{dt} + \omega_0^2(y) = -\frac{d^2x}{dt^2} = g(t)$$

where $\omega_0^2 = (2\pi f_0)^2$ is the resonant frequency, equal to k/m for an undamped mass m hanging on a spring of constant k. The ground acceleration is $g(t)$. The left-hand side is the equation of a damped harmonic oscillator, and the right-hand side is the vertical acceleration of Earth's surface as a forcing function. The damping is inversely proportional to a dimensionless parameter Q, the quality factor. Large Q means small damping and vice versa. In a geophone (see below), the output voltage is proportional to the velocity of the mass dy/dt, with respect to the surrounding coil, and a resistor shunted across the coil converts electric current to heat to provide damping.

The analog system transfer function is obtained by assuming all time functions are proportional to $\exp(i2\pi ft)$, taking derivatives, and rearranging terms. Alternatively, the derivative theorem for Fourier transforms (Appendix B) gives the same result. The transfer function $L(f)$ has a quadratic polynomial in f in its denominator, which can be factored as follows:

$$L(f) = \frac{Y(f)}{G(f)} = \frac{-1}{(2\pi)^2(f - f_+)(f - f_-)}$$

Here $Y(f)$ and $G(f)$ are the Fourier transforms of $y(t)$ and $g(t)$.

The complex roots of the denominator polynomial are

$$f_+ = \frac{if_0}{2Q} + \sqrt{f_0^2 - \left(\frac{f_0}{2Q}\right)^2}$$

$$f_- = \frac{if_0}{2Q} - \sqrt{f_0^2 - \left(\frac{f_0}{2Q}\right)^2}$$

Thus, the transfer function contains factors in the denominator (giving rise to poles) like those of a digital filter. But these poles are located within the complex-frequency plane, the f plane, rather than the complex Z plane.

As described in Appendix B, the inverse Fourier transform gives the seismometer mass displacement for a given ground acceleration $g(t)$ in terms of its Fourier transform $G(f)$:

$$y(t) = \int_{-\infty}^{+\infty} \frac{G(f)}{(2\pi)^2(f - f_+)(f - f_-)} \exp(2\pi i ft) \, df$$

which is an integral along the real axis in the complex-frequency plane. The integral shows how, at each frequency, the input ground acceleration Fourier transform $G(f)$ is magnified near the complex resonant frequencies f_+ and f_-. This behavior is similar to the relationship between Z plane pole locations and the unit circle. It suggests that we place poles in the Z plane to obtain a digital filter transfer function with similar behavior. That is, we need to map frequency-plane poles to Z plane poles. The matched Z transform is one mapping scheme, leading to a transfer function

$$\frac{1}{(Z - \exp(-2\pi i f_+))(Z - \exp(-2\pi i f_-))}$$

where we have mapped a point in the complex-frequency plane f_0 to a point in the complex Z plane:

$$f_0 \rightarrow Z = \exp(-2\pi i f_0)$$

When the frequency of a digital time series is close to f_0, the filter transfer function becomes large. Positive Q values result in Z plane poles outside the unit circle, as needed for a stable filter. Once we have the transfer function written as a ratio of Z polynomials, we can write down the digital filter equation.

There is a second method for mapping poles and zeros in the f plane to the Z plane, called a bilinear transformation:

$$f_0 \rightarrow \frac{1}{i\pi} \frac{(1-Z)}{(1+Z)} = \frac{1}{\pi} \tan(\pi f_0)$$

This also results in a stable filter. The bilinear transformation is one-to-one between the complex Z and f planes. That is, zero frequency maps to the point $Z = 1$ in the Z plane, while infinite frequency maps to the point $Z = -1$. Using the bilinear transform, every point in the complex f plane maps to a single point in the complex Z plane. In contrast, the matched Z transform wraps the infinite-frequency f axis around the unit circle, so that all alias frequencies map into the same locations on the unit circle. Both the bilinear and matched Z transforms may or may not yield a digital filter that resembles the analog transfer function in a satisfactory way over the range of frequencies of interest. In that case, one is free to add additional poles and zeros to tune the digital filter transfer function as needed.

Figure 8.5 shows the analog power transfer functions from the differential equation and the transfer functions of two digital filters derived using respectively matched Z and bilinear transforms. This is the particular case of a 12 Hz resonant frequency (typical of a short-period vertical seismometer, called a geophone), with $Q = 2$. In this case, the bilinear transform yields a closer match to the analog transfer function.

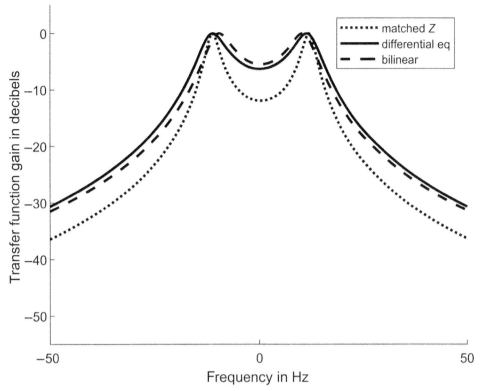

Figure 8.5 The power transfer function of a seismometer (simple harmonic oscillator) differential equation when $Q = 2$, $f_0 = 12$ Hz, shows a maximum near the 12 Hz resonant frequency. Digital filter power transfer functions obtained using matched Z and bilinear transforms are plotted with it. The digital sample interval is 2 milliseconds, with Nyquist frequency equal to 250 Hz. Better agreement with the differential equation transfer function is found using a bilinear transform. Both transforms are mappings of the complex f plane (associated with the linear differential equation describing the seismometer) to the complex Z plane (associated with the digital filter transfer functions). In these mappings, the real axis of the complex f plane is mapped onto the unit circle in the Z plane. The upper half of the f plane is mapped outside the unit circle in the Z plane. As shown in the text, the poles associated with the seismometer transfer function lie in the upper half of the f plane owing to damping (positive Q values). When mapped into the region outside the unit circle in the Z plane, the resulting digital filter is stable.

8.5 Derivative and Integration Filters

We now consider another application of mapping poles and zeros from the complex f plane to the Z plane to construct linear filters approximating differentiation and integration. The need to find the derivative or integral of a time series arises often in geophysics. For example, multiple seismic instruments may be installed at a single location to measure seismic waves, but each may measure a different property of the waves. Some seismic stations may include: a GPS receiver (described in Chapter 9) which measures displacement; a seismometer (similar to that in the previous section) measuring velocity (the time derivative of displacement); and an accelerometer (measuring the time derivative of velocity). Each instrument has distinct capabilities in terms of frequency response and dynamic range. For example, only a GPS receiver is able to measure displacements of meters over time scales of seconds and longer in the near field of an earthquake. Derivative and integration filters are required to combine time series from all instruments to obtain a complete picture of ground motion.

Finding the derivative of a continuous function corresponds to applying a linear filter d/dt. Its transfer function is $L(f)$ obtained by applying it to $\exp(i2\pi ft)$, finding that

$$L(f) = 2\pi i(f - 0)$$

where we write $(f - 0)$ to remind us that $L(f)$ has a zero in the complex frequency plane at $f = 0$. The matched Z transform can be used to find a digital filter by placing in the Z plane a zero at $Z = \exp(2\pi i0) = 1$. The filter transfer function is $(1 - Z)$, which correponds to the digital filter equation $y_t = x_t - x_{t-1}$, recognized as a first difference, an approximation to a derivative used in simple numerical finite difference schemes.

If the bilinear transformation is used in place of the matched Z, the transfer function is

$$\frac{2(1 - Z)}{1 + Z}$$

for which the digital filter is

$$y_t = 2x_t - 2x_{t-1} - y_{t-1}$$

This can also be written as

$$\frac{y_t + y_{t-1}}{2} = x_t - x_{t-1}$$

which is the Crank–Nicolson scheme used in finite-difference solutions of differential equations.

Integration is a linear operation that is inverse to taking a derivative. The inverse of the first-difference filter is

$$y_t = x_t + y_{t-1}$$

This filter accumulates the sum of x_t values, a simple form of integration when the sum is multiplied by the interval width Δt.

The inverse of the derivative filter via the bilinear transformation (interchanging MA and AR coefficients) is

$$y_t = (1/2)x_t + (1/2)(x_{t-1}) + y_{t-1}$$

This corresponds to trapezoidal rule integration, a numerical integration scheme superior to the accumulation of a sum.

8.6 Echo and Reverberation Filters

In reflection seismology, seismograms contain reverberations within the water layer (in marine seismology), and reflections from the free surface near the source or receiver (ghost reflections). The addition of these reflections to the directly arriving signal can be modeled as linear filters. The transfer functions of these filters show that frequencies are reduced or magnified. When formulated in the context of linear filters, our understanding of inverse filters can be used to remove these nuisances from seismograms. In the development here we assume a very simplified geometry and

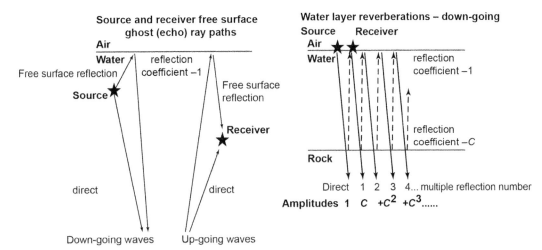

Figure 8.6 Schematic figure showing ray paths for source and receiver ghosts on the left, and for water-bottom multiple reflections on the right. Rays are associated with vertically propagating plane waves, and are therefore vertical although they are shown as slightly tilted for illustration purposes. The ghosts on the left are free surface reflections near the source and receiver, whose locations are indicated by stars. At both the source and receiver locations, direct arrivals and a free surface reflection are present. On the right, multiple reflections occur within the water layer. Again the source and receiver are indicated by stars, and for vertically traveling plane waves are at the same location although they are separated to improve clarity in the figure. The reverberation acts as a filter for waves traveling into the Earth below the water layer, and also filters them, after reflection from depth, on their way back to the surface. The total reverberation filter is therefore a cascade of two one-way reverberation filters.

plane vertically propagating waves. The geometry of free surface ghosts is shown on the left side of Figure 8.6, and the geometry of water layer reverberations on the right.

If a seismic source is detonated below the sea surface, the effective source is the sum of the direct downward wave and an upward traveling wave reflected at the free surface, delayed in time and scaled by the reflection coefficient of -1 at the free surface. Similarly if a hydrophone (receiver) is located below the surface, the recorded signal will be the sum of the direct arrival from below and the surface reflection from above. In each case, the surface reflection is called a "ghost". The term originally derives from the appearance of a television image on a cathode-ray-tube scanned display, where the direct image was trailed by a "ghost" image due to the delayed signal reflected off a nearby building or hillside.

The impulse response of a free surface ghost filter can be represented by the time series (for plane waves, and free surface reflection coefficient of -1), as

$$l_t = [1, 0, 0, \ldots, -1]$$

where the number of zeros corresponds to the round-trip travel time (in units of Δt from the source to the free surface and back. If the source depth is d and the seismic wave speed is V, then the transfer function of this MA filter is

$$L(f) = [1 - \exp(i2\pi f (2d/V))]$$

This transfer function goes to zero at certain frequencies (called ghost nulls), so the effect is serious because the ghost reflection causes a loss of information at these frequencies. Thus the depth of a source, which controls the ghost nulls, is an important element in designing seismic experiments.

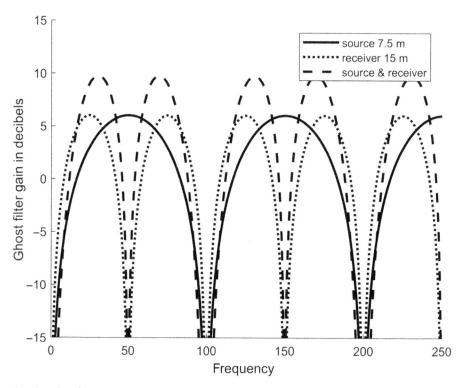

Figure 8.7 Source and receiver ghost filter transfer functions for a source depth of 7.5 meters and receiver depth of 15 meters. The cascade of the two is the combined ghost filter. The source and receiver depths are normally chosen so that ghost nulls, where the transfer function goes to zero, are outside the range of frequencies of interest. In addition, the source and receiver depths are normally chosen so that the null of each is at the same frequency.

Furthermore, an exact de-ghosting filter is not possible (for this simple model) because certain frequencies are lost. Figure 8.7 shows that when source and receiver are placed at two different depths (source at 7.5 m, receiver at 15 m) ghost nulls occur at different frequencies, causing a loss of signal at both 50 and 100 Hz. If both source and receiver were placed at 7.5 m depth then there would be no loss of signal at 50 Hz, which is in the middle of the frequency band commonly used in seismic exploration.

In marine seismic reflection profiling, the water-bottom reflection coefficient is large, due to the contrast in density and wave speed at the sea bottom. Seismic waves are strongly reflected both from the bottom and the sea surface, so many reverberations may occur. These water-bottom multiple reflections can obscure waves reflected from below the sea floor. If these reverberations can be described by a linear filter, then an inverse filter can be applied to remove them.

Suppose that the water-bottom reflection coefficient is c (a number with magnitude less than 1) and the free surface reflection coefficient is again taken to be -1. Refer to Figure 8.6, right-hand side, to understand the sequence of multiple reflections.

The two-way vertical travel time in the water layer is $T = 2d/V$ where V is about 1500 m/s and d is the water depth. A plane wave source (impulse) of magnitude 1 is located just below the surface, and sends waves downward. Just below the sea floor, ignoring transmission losses, the first arrival is the impulse (the direct arrival), followed by delayed multiple reflections, each spaced in time by T.

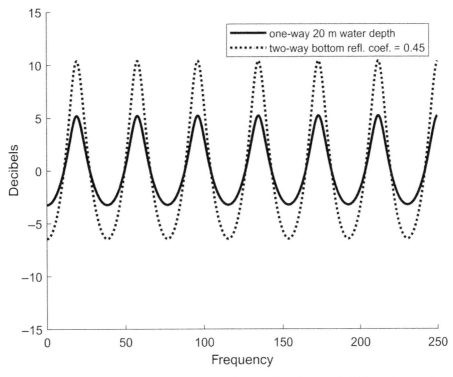

Figure 8.8 Power transfer function of one-way and two-way water layer reverberation filters, for water depth 20 m and a water-bottom reflection coefficient of 0.45. Seismic waves generated at the surface and reflected from depths below the water layer must pass through the reverberant region twice, so the effect is a cascade of two one-way filters. For a cascade of two filters, the gain in decibels is the sum of the gains from each filter. Therefore, the gain in decibels at the peaks and the attenuation at the troughs are each twice the number of decibels of the one-way filter gain.

The delay between the multiple reflections is $T = n\Delta t$, and the time series of events starts with the initial arrival of magnitude 1, followed by reverberations spaced by time T:

$$[1, 0, 0, \ldots, 0, 0, -c, 0, \ldots 0, c^2, 0, \ldots 0, -c^3, 0, \ldots]$$

There are $n - 1$ zeros between each event, and the time series is the one-way impulse response of the reverberating water layer, considered as a linear filter. It is infinite in length and decaying in time, suggesting that it may be regarded as the impulse response of an AR linear filter. In a seismic experiment, the waves of interest must pass through the layer twice, so the net filtering effect of the layer is the cascade of two one-way filters. The one-way filter is an AR filter with equation

$$y_t = x_t - c y_{t-n}$$

where n is the time value (in units of Δt) equal to T. The transfer function of both one-way and two-way reverberation filters is shown in Figure 8.8 for a typical water depth of 20 m. Within the typical range of frequencies used in seismic exploration, below 100 Hz, the effect of the reverberant layer is to strongly amplify the signal at several resonant frequencies (by over 10 dB in this example) while diminishing the signal at frequencies between resonant peaks.

In Section 5.7 it was shown that the inverse to a pure AR filter (here, the reverberation filter) is purely MA, so that the multiple reflections can be removed by twice-convolving the seismogram with the exact inverse:

$$l_t = [1, 0, 0, \ldots, c]$$

This inverse filter, called a Backus filter, provides one approach to suppressing strong water-bottom multiple reflections.

8.7 Sampled Impulse Response Filter Coefficients

If the continuous-time impulse response (IR) of a linear physical (analog) system is known, a linear digital filter approximation is obtained by sampling the impulse response. These samples are then used as coefficients of an MA filter. If the impulse response is infinite in duration, then it must be truncated in some way. Here we show that this is best done using a taper (a window function) to truncate in a gradual way.

Suppose we seek an MA filter to pass all frequencies from zero to half the Nyquist. The ideal transfer function is (from the Fourier transform theory in Appendix B) the boxcar function $II(2f)$, which is unity in the pass band, and zero in the stop band at frequencies above half Nyquist. The ideal filter is implemented for analog signals via convolution with the function $(1/2)\,\mathrm{sinc}(t/2)$ where the sinc function is $\mathrm{sinc}(t) = \sin(\pi t)/(\pi t)$. However, $(1/2)\,\mathrm{sinc}(t/2)$ is of infinite duration, so must be truncated. If we abruptly truncate it to 11 samples to form an MA filter the coefficients are

$$b = [0.0637, 0.0000, -0.1061, 0.0000, 0.3183, 0.5000, 0.3183, 0.0000, -0.1061, 0.0000, 0.0637]$$

If we taper these with a Hanning window whose values are

$$[0.0670, 0.2500, 0.5000, 0.7500, 0.9330, 1.0000, 0.9330, 0.7500, 0.5000, 0.2500, 0.0670]$$

then the windowed MA filter coefficients are

$$b1 = [0.0043, 0.0000, -0.0531, 0.0000, 0.2970, 0.5000, 0.2970, 0.0000, -0.0531, 0.0000, 0.0043]$$

Figure 8.9 compares the transfer functions of the truncated and Hanning tapered filters. The Hanning taper provides about 20 dB better rejection in the stop band, above half-Nyquist, with a slight change in the cutoff frequency, and a less steep slope near half the Nyquist frequency. This illustrates the usual trade-off between matching an ideal transfer function and reducing unwanted features, such as ripples in the transfer function stop band. To improve the filter, additional coefficients can be used, at the expense of additional computation. If low-pass filtering is the goal, allowing both AR and MA coefficients can be more efficient. The digital Butterworth filter whose coefficients are

$$b = [0.1667, 0.5000, 0.5000, 0.1667]$$

$$a = [1.0000, 0.0000, 0.3333, 0.0000]$$

is an example of an efficient ARMA low-pass filter. Its transfer function is shown for the same cutoff frequency (−3 dB point) of half-Nyquist. Butterworth filter design tools are widely available and provide filters with pass bands that are very flat, with very low side lobes in the stop band.

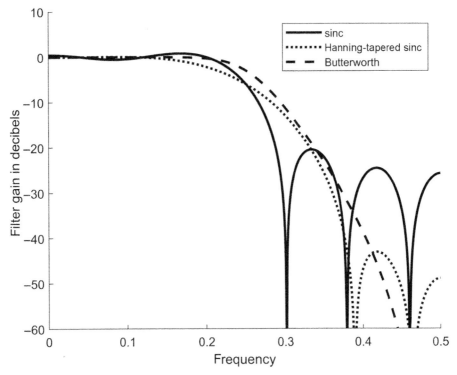

Transfer functions of low-pass filters designed to remove frequencies above half the Nyquist. An MA filter composed of sinc function values provides one choice for this filter. If sinc function is simply truncated to a finite length, the filter transfer function will then have relatively large side lobes in the stop band. This is not desirable; gradually tapering the function to zero using a Hanning window reduces the side lobes considerably. An alternative to a purely MA filter is to allow for autoregressive terms. A standard filter in this case is the Butterworth. The example here involves about the same amount of computation (since there is the same number of filter coefficients) but shows considerably better behavior in its transfer function, which is more nearly constant in the pass band, with lower side lobes in the stop band.

8.8 Gravity Anomaly Calculations

Gravity anomalies are measured at many spatial scales with applications to resource exploration, regional tectonics, and global mantle dynamics. Prediction of gravity anomalies at a spatial scale where the Earth is considered flat can be posed as a linear filter problem. The filter input is a sequence (linear array) of discretely sampled subsurface density variations, and the filter output is discretely sampled anomalous variations in the gravity field measured above the masses at or above the surface. Flat-Earth geometry is used in exploration and regional tectonic studies.

Gravity anomalies do not provide a unique measure of the subsurface because an infinite number of density variations may produce exactly or nearly the same gravity anomalies at the surface. Therefore, a traditional use of gravity observations is to test a hypothesis. In the process, one proposes a model of subsurface density variations, predicts the associated gravity anomalies, and then compares the prediction with the observed values. Additional steps may involve least squares adjustment to refine the proposed density model to better match observations.

Consider measuring the gravity anomaly at discrete locations along a surface profile using a survey (relative) gravimeter, which senses only changes in gravity. In practice the Earth is three-dimensional, but, to simplify the discussion, assume that gravity anomalies are measured along a single profile and are due to anomalous masses within a vertical plane containing the survey profile. Also, in practice, various contributions to gravity change must be removed from the observed changes to isolate the effects of local structures of interest. Typically one subtracts gravity effects of Earth tides, local topography, changes in latitude, variations in barometric pressure, and effects of deep structures (lower crust and upper mantle) that are not relevant. In concept, a survey gravimeter measures small changes in the stretch of a spring suspending a proof mass. The spring is aligned with the vertical direction, which is perpendicular to the flat observation surface. The spring stretches or contracts due to the changing weight of the proof mass as gravity varies with position. The actual design of survey gravimeters is more complex, and gravimeter design remains a subject of continuing research.

We will predict the gravity anomaly due to a spherical anomalous mass in the subsurface. A spherical mass has the same gravity field as a point mass located at its center. The anomalous mass is imbedded in a homogeneous half-space, as shown in Figure 8.10, and we will compute the anomaly for a mass at two different depths. To retain consistency with time series notation, the horizontal coordinate is t, while lower case z denotes depth below the observation surface $z = 0$. The Greek letter ρ refers to the anomalous density, so the anomalous mass at a given point is the product of the local value (on a discrete grid) of ρ times the local discrete volume element, which is taken to be unity. The anomalous density may be either positive or negative, since it is the difference between the local density and the homogeneous half-space density. The force of gravity acting on the proof mass is given by Newton's inverse square law. The universal gravitational constant is denoted by G. The force is directed along the radial line between each observation point at position t and the anomalous density at $t = 0$ and depth z. The measured gravity anomaly is the vertical component of this force because this is the direction in which the spring in the gravimeter is being stretched. The vertical component is the force along the radial direction given by the inverse square law multiplied by the cosine of the angle between the vertical and radial direction

$$g_t = \frac{G\rho}{(t^2 + z^2)} \times \frac{z}{(t^2 + z^2)^{1/2}}$$

The resulting gravity anomaly is symmetric about the location of the mass at $t = 0$.

$$g_t = \frac{G\rho z}{(t^2 + z^2)^{3/2}}$$

This is the impulse (point mass) response of a linear filter $l_t = g_t$ when $\rho = 1$ in the previous equation. That is, the sequence of discrete samples of density along the t axis is $\rho_t = \delta_t$ with values equal to zero except for the unit density at $t = 0$. The gravity field of several masses is just the sum of their individual fields, which allows us to predict the gravity anomaly along the profile due to a sequence of anomalous masses ρ_t at discrete locations t, at the same depth z beneath the surface. This is just the convolution $l_t * \rho_t$, the superposition of anomalies from all masses at depth z. We can obtain a complete gravity anomaly by considering anomalous masses at different depths. The impulse response is different for each depth, as shown by the different shapes in Figure 8.10. However, the contribution from each depth is a convolution with the impulse response for that depth, so the total anomaly is found by summing effects from all depths. In a realistic application, the

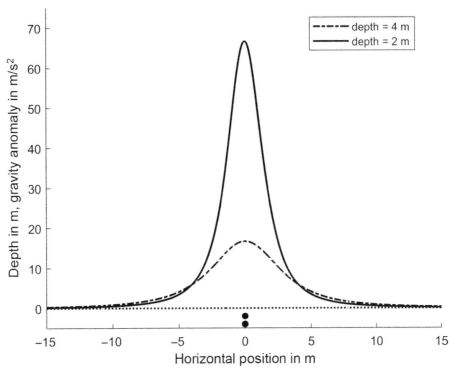

The gravity anomalies along a horizontal profile (the t-axis) at the surface (where depth $z = 0$) due to anomalous point masses at depths of $z = 2$ or 4 meters. These anomalies are proportional to the linear filter impulse response for the respective depths. The mass locations are identified by black circles at $t = 0$ and $z = 2$ and 4 meters. The anomalous mass is 4900 kg, the product of 4900 kg/m^3 times a one-cubic-meter volume element. An anomalous mass of 4900 kg corresponds, for example, to a one-cubic-meter iron sphere (density 7900 kg/m^3) imbedded in homogeous sediments of density 3000 kg/m^3. The gravity anomalies are plotted on the vertical scale in acceleration units of nm/s^2. Recognizing that the ambient gravity field at Earth's surface is about 9.8 m/s^2, one nm/s^2 is a very small local change in gravity, about 1 part in 10^{10}. The precision of the best survey gravimeters is near 50 nm/s^2, so the anomaly due to the mass at 2 meters depth would be observable. Notice that the 2 meter anomaly is both larger in magnitude and narrower in width, implying increased high-spatial-frequency content. Closer proximity to the anomalous mass at 2 meters depth yields both a higher amplitude and better spatial resolution (associated with higher-spatial-frequency content), to allow a more precise determination of mass location. As described in the text, it is possible to apply a linear filter to downward-continue surface observations of the 4 meter anomaly to improve spatial resolution. A 2 meter downward-continuation filter applied to the 4 meter anomaly would change its appearance to that of the 2 meter anomaly. Downward continuation is possible if there are no anomalous masses between the observation and continuation elevations. A downward-continuation filter amplifies high spatial frequencies. If one attempts to downward-continue over too great a distance, the noise or random errors along the profile will be magnified and will obscure the gravity signal.

three-dimensional nature of the subsurface must be considered, requiring a two-dimensional impulse response (depending on both t and the coordinate normal to the (t, z)-plane). Then, a two-dimensional convolution over a planar grid of mass anomalies at each depth z would be required, followed by the additionn of contributions from all depths.

In practice, this method of predicting gravity anomalies is rarely used because it is computationally inefficient. Efficient algorithms make use of the Discrete Fourier Transform. Other methods approximate density variations in the subsurface using prismatic bodies for which anomalies can be calculated exactly. Another issue with the spatial convolution approach is that the impulse response extends to infinite values of horizontal coordinate t. So, computations require that after sampling at discrete values of t, the impulse response must be truncated to finite length using a window function, as described in the previous section.

Nevertheless, recognizing that the calculation of gravity anomalies can be posed as a linear filtering problem provides insight into an important aspect of gravity data processing: upward and downward continuation. A continuation filter transforms a gravity anomaly profile observed at one elevation so that it appears to have been observed at another. Continuation filtering is possible providing there are no mass anomalies between observation and continuation elevations. A downward-continuation filter transforms the anomaly to an elevation closer to the source, while an upward-continuation filter does the opposite. An example application is the downward continuation of airborne gravity surveys flown at high altitude for safety reasons. Figure 8.10 shows that downward continuation has two desirable effects. If the observation elevation is closer to the anomalous mass then the anomaly amplitude is increased and spatial resolution is improved. Considering the two impulse responses in Figure 8.10, a 2 meter downward continuation filter would be a zero-phase MA filter that converts the impulse response for 4 meters depth into that for 2 meters depth. Coefficients for such a filter can be developed from the theory of gravity fields, or via least squares as discussed in the next chapter. In practice, however, continuation filters are commonly implemented using the Discrete Fourier Transform. By comparing the impulse responses in Figure 8.10 it is evident that upward continuation reduces the amplitudes of high spatial frequencies while downward continuation amplifies them. Observed gravity profiles inevitably contain errors, so the application of a downward continuation filter is generally limited by the level of high-spatial-frequency noise.

8.9 Ground Motion Amplification in an Earthquake

Very large surface displacements are observed in some earthquake-prone regions, often leading to destruction of buildings and other structures. Several physical processes may be responsible for very large ground motion. An important one is liquefaction, in which water-saturated granular soils lose all shear strength when strongly shaken by seismic waves. Liquefaction is not a linear process, because it is highly dependent on the amplitude of shaking, but another important mechanism, amplification due to soft near-surface soils, can be described using a linear filter model.

The transfer function describing the amplifying effect of soft near-surface material depends mainly on the near-surface variations in shear strength. These are determined by measuring vertical variations in the seismic shear wave speed. The transfer function computed for local site conditions will identify frequencies that will be strongly amplified, and if these are near resonant frequencies of proposed or existing structures, then damage or destruction may result. The transfer function will then guide preventative measures in site preparation, building design, and building retrofitting. By adopting a linear filter model, it is assumed that physical properties near the surface are the same for all wave amplitudes. In fact, very large amplitude seismic waves may experience reduced shear strength and increased damping, but the linear filter description can be modified to include these effects.

Consider a single horizontal layer having a low shear wave speed over a half-space of higher speed. We will consider vertically propagating shear waves (horizontally polarized SH waves arriving from below). The low-speed surface layer 1 lies over a higher-speed infinite half-space (layer 2). Layer 1 has thickness h, shear wave speed Vs_1 and density ρ_1. The half-space has wave speed Vs_2 and

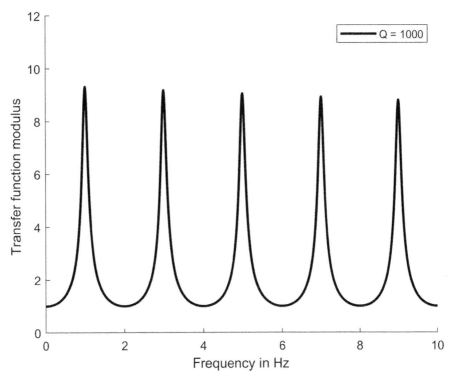

Figure 8.11 The linear filter transfer function modulus gives the ratio of surface displacement relative to that at the layer (1–2) interface for the case of SH waves arriving from below. Layer 1 is a soft soil layer over a half-space. Parameters for the Earth model are $\rho_1 = 2000\,\text{kg/m}^3$, and $\rho_2 = 2500\,\text{kg/m}^3$, with shear velocities $Vs_1 = 200$ m/s and $Vs_2 = 1500$ m/s. The transfer function has been evaluated for complex frequencies, as described in the text, to allow for a small amount of damping ($Q = 1000$). As a result of the damping, there is a slight decrease in transfer function amplitude with increasing frequency. With large amounts of damping (low Q), only the lowest-frequency peaks are important. In a practical application, the vertical shear-speed variations at a site would be measured and the transfer function evaluated in order to develop building designs resistant to shaking at the frequencies of the transfer function peaks.

density ρ_2. The transfer function is found by considering sinusoidal SH waves of frequency f in the half-space below, imposing boundary conditions of displacement continuity at the layer (1–2) interface, and zero stress at the surface. If we define I to be the shear wave impedance ratio as

$$I = \frac{\rho_1 V s_1}{\rho_2 V s_2}$$

the transfer function is

$$L(f) = \frac{1}{[\cos(2\pi f h/V s_1) - iI \sin(2\pi f h/V s_1)]}$$

A complex transfer function implies a phase shift between the maximum amplitude at the bottom of layer 1, and the surface. Assume a 50 m thick soil layer 1, where $\rho_1 = 2000\,\text{kg/m}^3$. For the half-space (2) $\rho_2 = 2500$. Shear wave speeds are $V s_1 = 200$ m/s and $V s_2 = 1500$ m/s. The modulus of the complex transfer function is shown in Figure 8.11. It has a series of peaks at resonant frequencies within the near-surface layer. The peak amplification is nearly an order of magnitude for the particular parameters here. The first several peaks at low frequencies are the most important because attenuation at higher frequencies (described by the quality factor Q) typically diminishes their amplitude. Figure 8.11 was calculated for a very-low-attenuation case ($Q = 1000$), so there is only a slight decrease in amplitude for the higher-frequency peaks. Exercise 8.3 shows that a more typical low Q value for soft sediments has a much greater effect on the higher-frequency peaks.

8.10 Chapter Summary

This chapter has provided examples of linear filter design methods for: data processing purposes (notch, derivative, and integration filters); for simulating physical systems described by linear differential equations; and for describing physical processes. The main points are as follows.

- The pole and zero locations relative to the unit circle in the Z plane control transfer function gain, filter stability, and existence of a stable inverse.
- Continuous-time linear systems described by differential equations or linear operations such as integration or differentiation can be approximated using digital filters. Two mapping schemes (matched Z and bilinear transform) are useful for mapping poles and zeros from the complex frequency plane associated with continuous time transfer functions to the complex Z plane of digital filter transfer functions.
- The impulse response of a continuous-time linear system can be sampled to provide MA filter coefficients. If the impulse response is infinite in duration, it must be truncated to finite length. This is best done by gradually tapering to zero, using a window function.
- Linear filters provide useful models for physical systems such as a seismometer, and for descriptions of physical processes such as the echoes and reverberations of seismic or acoustic waves, the prediction of gravity anomalies, and the amplification of near-surface ground motion in an earthquake.

Exercises

8.1 **Ghost Filters.** Refer to the discussion of echo (ghost) and reverberation filters. Assume the speed of sound in water is 1500 m/s, and the Nyquist frequency is 500 Hz.

  A. Suppose that an airgun source has a peak output at 30 Hz, with very little signal above 60 Hz. What depth should the airgun be placed to make best use of the ghost filter transfer function peak and trough?

  B. The total ghost filtering effect is the cascade of source ghost filter and receiver ghost filter. Plot the power transfer functions of the cascade of the two ghost filters for the cases when they are at the same depth as found in the previous part A, and when receiver depth is 50 percent greater than source. Explain which of these choices might be better than the other.

8.2 **Water Layer Reverberation Filters.** Assume the Nyquist frequency is 500 Hz and speed of sound in water is 1500 m/s. Find the two-way reverberation filter equation for a cascade of two one-way filters, and plot its transfer function for a water depth of $d = 7.5$ meters. Calculate the reflection coefficient at the water bottom, if the rock velocity is 2000 m/s and the rock density is 2000 kg per cubic meter. (The reflection coefficient c is the difference in the values of the quantity velocity times density across the water-bottom interface, divided by the sum of these values.) Find the filter that removes the multiple reflections, and plot its transfer function gain on the same plot.

8.3   **Ground Motion Amplification.** Allow for damping in the near-surface layer described in the Section 8.9 by introducing a complex frequency as in Chapter 3, where Exercise 3.2 shows how complex elastic parameters or a complex frequency may be used to describe damping. In addition, Section 8.4 shows how damping appears in a seismometer transfer function as a complex frequency.

Use the soft-soil parameters given in the text for a layer thickness 50 m, shear wave speeds of 200 and 1500 m/s, and densities of 2000 and 2500 kg/m^3. Plot the transfer function modulus of ground motion amplification for frequencies up to 10 Hz for two Q values (10 and 100), confirming that amplification by nearly an order of magnitude (approaching 20 dB) might occur at a frequencies below 5 Hz, a typical range for building resonant frequencies. You should also note that more damping (lower Q) diminishes the amplification at all frequencies and especially at higher frequencies. You may wish to experiment with the choice of a complex frequency $f(1-i/2Q)$, reversing the sign of the imaginary part. This will result in anti-damping, that is, reversing the sign will cause higher frequencies to be amplified. This is not physical, so it is important to establish conventions for stable damped systems by defining the imaginary part of frequency with the correct sign.

8.4   **Seeking a Stable Inverse Filter.** The 1999 Hector Mines data (Figure 1.6) are reported with a source time function, representing an effective wavelet l_t convolved with the seismogram of an impulsive source. That is, the observed seismogram is given by $y_t = l_t * x_t$, where y_t is one of the seismograms in Figure 1.6 and x_t is the impulsive source seismogram to be estimated. That is, we wish to find an inverse filter for l_t. The source time function l_t is 40 seconds long, given at one second intervals, in units of 10^{17} Nm/s, and appears in the file containing the seismogram values. Remove the leading and trailing zeros and set the mean value to zero. Normalize the amplitude by dividing all values by the first non-zero value. Then show that the source time function is not minimum-phase by confirming that its exact inverse filter (an AR filter) has an impulse response that grows exponentially. Exercise 9.4 shows how to find a stable inverse using least squares, and how to find a minimum-phase version of the source time function.

9

Least Squares and Correlation Filters

This chapter surveys least squares and related methods for designing inverse, prediction, and interpolation filters. The name Wiener filter is associated with these methods. Correlations and autocorrelations play a central role in forming normal equations for filter coefficients. The underlying theory assumes that true correlations are available, but in practice correlations and autocorrelations are estimated from data. We will find the surprising result that a least squares inverse to a linear filter with impulse response l_t can be found from its autocorrelation, without knowledge of the impulse response itself. The least squares inverse filter is closely related to a filter for predicting future values of a time series. Similar ideas may be used to design interpolation filters and to develop linear filter models for time series. This chapter also reviews the correlation filters used in geophysical systems such as radar, sonar, and exploration seismology. We also discuss how correlation filtering is the essential element enabling navigation with the Global Positioning System (GPS).

9.1 Least Squares Inverse Filters

Suppose that $y_t = l_t * x_t$ is a linearly filtered version of x_t and the l_t is known. Then y_t filtered by the exact inverse of l_t should yield x_t. The exact inverse l_t^{-1} will therefore have the property that

$$l_t * l_t^{-1} = \delta_t = [1, 0, 0, \ldots]$$

However, it is possible that the exact inverse of l_t has an infinite impulse response or is unstable, or that its transfer function is infinite at some frequencies. We can use least squares to find an approximate MA inverse filter with none of these problems.

Writing the convolution as a matrix multiplication gives the problem the form of a set of least squares observation equations. As a simple example, let l_t be of length 3, with first value $l_0 = 1$. To find its approximate least squares inverse p_t (also of length 3) such that $l_t * p_t \approx [1, 0, 0, \ldots]$ we use this matrix multiplication to represent convolution, as described in Section 5.4. The convolution matrix has five rows because the convolution produces a time series of length 5, the sum of the two lengths less 1. This leads to the observation equations (Section 7.3)

$$\begin{bmatrix} l_0 & 0 & 0 \\ l_1 & l_0 & 0 \\ l_2 & l_1 & l_0 \\ 0 & l_2 & l_1 \\ 0 & 0 & l_2 \end{bmatrix} \begin{bmatrix} p_0 \\ p_1 \\ p_2 \end{bmatrix} \approx \begin{bmatrix} 1 \\ 0 \\ 0 \\ 0 \\ 0 \end{bmatrix}$$

Multiplying both sides by the transpose of the convolution matrix yields normal equations with matrix elements containing sums over time of products $l_t l_{t-k}$, which are the autocorrelation values r_k of l_t at lag k:

$$\begin{bmatrix} \sum l_t^2 & \sum l_t l_{t-1} & \sum l_t l_{t-2} \\ \sum l_t l_{t-1} & \sum l_t^2 & \sum l_t l_{t-1} \\ \sum l_t l_{t-2} & \sum l_t l_{t-1} & \sum l_t^2 \end{bmatrix} = \begin{bmatrix} r_0 & r_1 & r_2 \\ r_1 & r_0 & r_1 \\ r_2 & r_1 & r_0 \end{bmatrix}$$

The matrix is symmetric and Toeplitz (having the same values along each diagonal), and is therefore suitable for solution by an efficient numerical scheme (the Levinson–Durbin algorithm):

$$\begin{bmatrix} r_0 & r_1 & r_2 \\ r_1 & r_0 & r_1 \\ r_2 & r_1 & r_0 \end{bmatrix} \begin{bmatrix} p_0 \\ p_1 \\ p_2 \end{bmatrix} = \begin{bmatrix} 1 \\ 0 \\ 0 \end{bmatrix}$$

Usually y_t is contaminated by noise n_t. This also should be considered, so that convolution with p_t does not amplify frequencies dominated by noise. The model of y_t is now

$$y_t = l_t * x_t + n_t$$

and the least squares inverse filtered p_t is designed so that

$$p_t * y_t = p_t * [l_t * x_t + n_t] \approx x_t$$

If the noise and signal ($l_t * x_t$) are uncorrelated, the normal equation matrix is the sum of the autocorrelation matrices of l_t and n_t. Usually we assume the noise is white, and, as demonstrated in Figure 2.3, the autocorrelation of white noise is effectively zero except at zero lag. As a result, the white noise autocorrelation matrix is diagonal, with zero off-diagonal elements and diagonal elements equal to the noise variance, which is taken to be constant in time. Thus, the normal equations are similar to the noise-free case but the diagonal elements are inflated by the noise variance σ_n^2. Using the length-3 example, the normal equations for p_t are given by

$$\begin{bmatrix} r_0 + \sigma_n^2 & r_1 & r_2 \\ r_1 & r_0 + \sigma_n^2 & r_1 \\ r_2 & r_1 & r_0 + \sigma_n^2 \end{bmatrix} \begin{bmatrix} p_0 \\ p_1 \\ p_2 \end{bmatrix} = \begin{bmatrix} 1 \\ 0 \\ 0 \end{bmatrix}$$

Because the noise variance may be unknown, it becomes an adjustable parameter in the filter design process. This is a form of damped least squares, an approach to regularizing least squares estimation problems. Increasing the assumed noise variance (called adding false white noise) reduces the amplification at frequencies where signal is small, making the action of the approximate inverse filter less aggressive. Reducing the order of the model (the length of p_t) has a similar effect, so both filter order and assumed noise level are adjustable parameters in the design of inverse filters.

9.2 Yule–Walker Equations

The Yule–Walker equations are used to obtain the coefficients of a linear filter that predicts the next value in a time series. The underlying assumption is that the time series values are random variables that are at least partially predictable from past values. This approach can be applied to other problems such as interpolation, developed later in this chapter. Surprisingly, we will find that an extended version of the Yule–Walker equations is identical in form to the equations developed in the previous section for finding the least squares inverse of a linear filter.

It is instructive to see how ordinary least squares might be used to solve the next-sample prediction problem. Suppose that y_t is partially predictable from a linear combination of its past values. If true, this would be immensely valuable in a variety of problems, such as forecasting weather, where physical processes are at work, or in economics (stock or commodity prices), where physics is absent but where empirical success might be possible.

We can use past observations in a time series y_t to train our prediction filter. We assume that N values $[y_0, y_1, \ldots, y_{N-1}]$ of a zero-mean time series are available and that the past two values will be used for prediction. The least squares observation equations for the prediction coefficients are given by

$$
\begin{bmatrix}
y_1 & y_0 \\
y_2 & y_1 \\
y_3 & y_2 \\
y_4 & y_3 \\
\cdots & \cdots \\
y_{N-2} & y_{N-3}
\end{bmatrix}
\begin{bmatrix}
-a_1 \\
-a_2
\end{bmatrix}
\approx
\begin{bmatrix}
y_2 \\
y_3 \\
y_4 \\
y_5 \\
\cdots \\
y_{N-1}
\end{bmatrix}
$$

where $[-a_1, -a_2]$ are the model parameters, and the use of negative signs follows the convention for AR linear filter coefficients from Chapter 5. This confirms that we are using least squares to develop an AR linear filter model of the form

$$
y_t = b_0 x_t - a_1 y_{t-1} - a_2 y_{t-2}
$$

Here, we do not know the time series x_t, but in other cases x_t, and other inputs, might be available and so could be included in the model. Time series models like this may include both moving average and autoregressive terms, and the development of such models is a standard problem in time series modeling as an application of least squares.

Considering the AR model in this case, the assumption is that each value of y_t is composed of a predictable part $-a_1 y_{t-1} - a_2 y_{t-2}$ and an unpredictable part $b_0 x_t$. The unpredictable part plays a role similar to the data errors in ordinary least squares. Multiplying the left- and right-hand sides by the matrix transpose, we find in the normal equation matrix sums of products that approximate autocorrelation values, with lag zero on the diagonal, lag 1 on the off-diagonals, and so on. As more data are added to the training set, the values in the normal equation matrix converge towards the autocorrelations appearing in the Yule–Walker equations.

The Yule–Walker equations address the problem of predicting the next value in a time series from a linear combination of past values. We illustrate the idea for predictions based on the past two values, as before. However, now we adopt a stochastic point of view, taking the time series values to be random variables, where expectations can be used to obtain the normal equations. Although the filter order is limited to 2 in this example, in practice many different orders can be tried to see which works best, in the sense of significantly improving prediction relative to a filter of lower order.

By using the expectation operator $E[]$, (described in Appendix C), we are assuming that both y_t and x_t are zero-mean random variables. Their associated pdf may be unknown, but the linear expectation operator allows us to introduce autocorrelation values such as $r_1 = E[y_t y_{t-1}]$. In practice these are estimated from samples of y_t in the usual way. The random variable x_t is taken to be the unit-variance zero-mean white noise, with zero autocorrelation at all lags except lag $k = 0$, where $E[x_t^2] = 1$. With this assumption, the variance of the unpredictable part of y_t is b_0^2.

We proceed by multiplying the model equation $y_t = b_0 x_t - a_1 y_{t-1} - a_2 y_{t-2}$ by y_{t-1}, then taking the expected value of both sides:

$$E[y_t y_{t-1}] = b_0 E[x_t y_{t-1}] - a_1 E[y_{t-1}^2] - a_2 E[y_{t-1} y_{t-2}]$$

The term on the left-hand side is the autocorrelation of y_t at lag 1, r_1. The first term on the right-hand side is zero, because x_t is uncorrelated with anything in the past, including y_{t-1}. The other terms are respectively autocorrelations at lags 0, 1, and 2. We can find one more equation on multiplying by y_{t-2}, using similar arguments to eliminate terms, and we obtain finally the Yule–Walker equations:

$$\begin{bmatrix} r_0 & r_1 \\ r_1 & r_0 \end{bmatrix} \begin{bmatrix} -\hat{a}_1 \\ -\hat{a}_2 \end{bmatrix} = \begin{bmatrix} r_1 \\ r_2 \end{bmatrix}$$

As we increase the filter order the matrix remains symmetric and Toeplitz with autocorrelation values of increasing lag away from the diagonal.

The Yule–Walker equations can be adapted to gapped prediction, that is, to predict over a gap of K samples. For example, an estimate of y_t that is K samples ahead of the previous two values uses coefficients that are solutions to a set of normal equations that, on the left, is the Toeplitz matrix of autocorrelations between what we have available (the data to be used in the estimate), and, on the right-hand side, contains cross correlations between what we have available and what we want to predict. To predict K samples ahead, the left-hand side remains the same (the same data are used to make the prediction) but the right-hand side becomes the autocorrelation values at lags K and $K + 1$, in this example:

$$\begin{bmatrix} r_0 & r_1 \\ r_1 & r_0 \end{bmatrix} \begin{bmatrix} -\hat{a}_1 \\ -\hat{a}_2 \end{bmatrix} = \begin{bmatrix} r_K \\ r_{K+1} \end{bmatrix}$$

So $K = 1$ gives the usual Yule–Walker equations, and $K > 1$ yields a gapped prediction filter. The latter is more efficient, for a given number of coefficients, if there is additional information concerning a delay time in the original AR model.

Other algorithms are available to estimate one-sample-forward prediction coefficients. Among these is the Burg algorithm, developed in the late 1960s to provide estimates without requiring the intermediate step of computing autocorrelation values. The Burg algorithm provides one-sample-forward prediction but is not easily adapted to gapped prediction, as were the Yule–Walker equations. This makes it useful to have a general method to obtain K-sample-forward coefficients from one-sample-forward coefficients $[a_1, a_2]$. To find gapped prediction coefficients for this general case, we introduce some new notation, using upper case letters to denote vectors and matrices (no connection with the frequency domain is implied). Define the state vector at time t, Y_t, as

$$Y_t = \begin{bmatrix} y_t \\ y_{t-1} \end{bmatrix}$$

and the state vector at time $t - 1$ as

$$Y_{t-1} = \begin{bmatrix} y_{t-1} \\ y_{t-2} \end{bmatrix}$$

Then the AR model relates the two by

$$y_t = -a_1 y_{t-1} - a_2 y_{t-2}$$

as a one-step-forward prediction from time $t-1$ to t. The state vector matrix notation for the same equation is

$$Y_t = AY_{t-1}$$

where the first row of matrix A is $[-a_1, -a_2]$ and the second row is $[1, 0]$. Then, propagating K steps forward requires multiplying by matrix A repeatedly, so that the K-sample-forward prediction involves higher powers of A:

$$Y_{t+K-1} = A^K Y_{t-1}$$

Thus the coefficients for the K-sample-forward prediction appear on the first row of the matrix A^K. For AR order m, the first row of matrix A is the set of AR coefficients $[-a_1, -a_2, \ldots, -a_m]$ and the lower part of A is the $m \times m$ identity matrix without its last row.

9.3 Interpolation Filters

Interpolation by standard methods (cubic spline, linear, and others) provides useful results but no measure of the possible interpolation errors. Discrete Fourier Transform interpolation provides band-limited results but requires uniformly sampled data. However, the interpolation of irregularly spaced data can be accomplished using methods similar to the Yule–Walker equations that also give an estimate of the interpolation error. The approach is commonly used in the spatial interpolation of geophysical data, providing both a useful interpolated value and a measure of error. This approach to interpolation is variously called kriging or geostatistics (in geology and geophysics), optimal interpolation (in meteorology), or least squares co-location (in geodesy). In some applications a unique terminology has evolved. For example, in kriging the term semi-variogram refers to zero-lag autocorrelation minus that at another lag. Regardless of the terminology, the key element is to make use of correlations between the available data and between the available data and the location where an estimate is sought. The dependence of correlation dependence on lag (spatial separation) is typically estimated from the available data. If few data are available, a functional form of correlation dependence on separation may be assumed. For example, linear, exponential, or other functions may be used to describe a decrease in correlation with increasing separation. Once the spatial correlation (that is, autocorrelation) properties are set, the method may be applied to data that are regularly or irregularly sampled in multiple spatial dimensions.

To see how this works and how to generate an estimate of interpolation error, we will use time series interpolation as an example. Suppose we have a time series x_t sampled at integer times, and we wish to find a linear filter that performs midpoint interpolation, using a linear combination of the nearest six values, three on each side. Taking the interpolation point to be at time $t + 1/2$ we minimize the expected value of the squared error,

$$E[(x_{t+1/2} - \hat{x}_{t+1/2})^2]$$

where the quantities are treated as zero-mean random variables:

$$\hat{x}_{t+1/2} = a_1 x_{t-2} + a_2 x_{t-1} + a_3 x_t + a_4 x_{t+1} + a_5 x_{t+2} + a_6 x_{t+3}$$

As in the derivation of the Yule–Walker equations, we multiply through successively by $x_{t-2}, \ldots, x_{t+3}$, and take the expectation of each value. The linear normal equations for the unknown coefficients are then given by

$$
\begin{bmatrix}
r_0 & r_1 & r_2 & r_3 & r_4 & r_5 \\
r_1 & r_0 & r_1 & r_2 & r_3 & r_4 \\
r_2 & r_1 & r_0 & r_1 & r_2 & r_3 \\
r_3 & r_2 & r_1 & r_0 & r_1 & r_2 \\
r_4 & r_3 & r_2 & r_1 & r_0 & r_1 \\
r_5 & r_4 & r_3 & r_2 & r_1 & r_0
\end{bmatrix}
\begin{bmatrix}
\hat{a}_1 \\ \hat{a}_2 \\ \hat{a}_3 \\ \hat{a}_4 \\ \hat{a}_5 \\ \hat{a}_6
\end{bmatrix}
=
\begin{bmatrix}
r_{5/2} \\ r_{3/2} \\ r_{1/2} \\ r_{1/2} \\ r_{3/2} \\ r_{5/2}
\end{bmatrix}
$$

The matrix on the left contains the autocorrelations for the available data. In this case it is Toeplitz because the time series is uniformly sampled. In the first row is the autocorrelation of x_{t-2} with each of the six data, starting with itself in column 1. The column vector on the right contains the correlations between what we have available (the six surrounding data) and what we want to estimate (the value at the midpoint). These correlations on the right are at non-integer lags, for example, $r_{1/2} = r_x(1/2)$. These could be estimated for integer lags from the actual data and then interpolated in some way. Alternatively, one could fit a functional form for correlation as a function of separation. Here we use the relationship between the autocorrelation and the spectrum, from Fourier transform theory, to determine autocorrelation values.

To proceed with this example we will use results concerning the Fourier transforms of continuous functions developed in Appendix B. Suppose we take x_t to be samples of a continuous function $x(t)$ whose spectrum is constant (white) with a value of unity up to the Nyquist frequency. The spectrum is therefore a boxcar function, equal to unity in the Nyquist interval $[-1/2, 1/2]$ and zero elsewhere. Because there are no frequencies outside the Nyquist band, the sampling theorem is followed by sampling $x(t)$ at integer time to obtain the time series x_t. From the autocorrelation theorem for Fourier transforms, which is analogous to that for discrete time series, discussed in Section 6.3, the autocorrelation of x_t is the Fourier transform of the boxcar function, equal to $\operatorname{sinc}(\tau)$, as discussed in Appendix B. Values of the sinc function appear on both left- and right-hand sides of the normal equations. The sinc function is zero at integer lags, and equal to unity at zero lag. This makes the left-hand matrix equal to the identity matrix. On the right-hand side are values of the sinc function at lags 0.5, 1.5, and 2.5. These are the interpolation coefficients:

$$[a_1, \ldots, a_6] = [0.1273, -0.2122, 0.6366, 0.6366, -0.2122, 0.1273]$$

These are precisely the weights (interpolation coefficients) in Whittaker's interpolation formula, described in the sampling theorem proof in Appendix B. If we know that x_t is smoother than this then the coefficients are different. For example, if the spectrum is flat but goes to zero beyond half the Nyquist, the autocorrelation function is $(1/2)\operatorname{sinc}(\tau/2)$ and the left-hand matrix is no longer the identity matrix. The solution is then

$$[a_1, \ldots, a_6] = [0.0238, -0.1244, 0.6016, 0.6016, -0.1244, 0.0238]$$

The sum of the interpolation coefficients is 1.0019, close to unity, and we can make the sum exactly unity by appropriate normalization if desired. In either case, reasonable results are obtained.

To get an estimate of the interpolation error, autocorrelation and cross correlation values are used. To simplify the discussion, let R represent the autocorrelation matrix, q be the right-hand side column vector, a be the column vector of estimated coefficients, and d be the column vector of the available data. Then the normal equations are given by $Ra = q$, so that $a = R^{-1}q$ and the estimate at the midpoint is $\hat{x}_{1/2} = a^T d$, where the superscript T denotes matrix transpose. The autocorrelation matrix $R = E[dd^T]$ gives the expected values of the correlations among the available data.

The expected squared error of the estimate is then

$$\sigma^2(\hat{x}_{1/2}) = E[(x_{1/2} - \hat{x}_{1/2})^2] = E[x_{1/2}^2 - 2q^T R^{-1}q + q^T R^{-1}dd^T R^{-1}q]$$

Substituting $R = E[dd^T]$ and r_0 for $E[x_{1/2}^2]$ the result simplifies to

$$\sigma^2(\hat{x}_{1/2}) = r_o - q^T R^{-1}q$$

To see that this is a sensible answer, consider the case when the values in q (the cross correlations between the available data and the quantity to be estimated) are zero. Then the values of a are zero, and the error variance is the zero-lag value r_0. Otherwise, the error variance is r_0 reduced by information provided from samples at surrounding points.

9.4 Prediction Error Filters

Returning again to the Yule–Walker equations for an order-2 model, one more equation is obtained by multiplying y_t by itself and finding the expected values:

$$E[y_t y_t] = r_0 = b_0^2 - a_1 r_1 - a_2 r_2$$

This can be combined with the previous two Yule–Walker equations in a set of normal equations in three unknowns b_0, a_1, and a_2:

$$\begin{bmatrix} r_0 & r_1 & r_2 \\ r_1 & r_0 & r_1 \\ r_2 & r_1 & r_0 \end{bmatrix} \begin{bmatrix} 1 \\ \hat{a}_1 \\ \hat{a}_2 \end{bmatrix} = \begin{bmatrix} b_0^2 \\ 0 \\ 0 \end{bmatrix}$$

This provides equations for the AR coefficients and white noise variance b_0^2 when rearranged as follows:

$$\begin{bmatrix} r_0 & r_1 & r_2 \\ r_1 & r_0 & r_1 \\ r_2 & r_1 & r_0 \end{bmatrix} \begin{bmatrix} 1/\hat{b}_0^2 \\ \hat{a_1/b_0^2} \\ \hat{a_2/b_0^2} \end{bmatrix} = \begin{bmatrix} 1 \\ 0 \\ 0 \end{bmatrix}$$

These equations are of exactly the same form as those for the least squares inverse filter p_t, showing the close connection between prediction and inverse filters. Multiplying the solution of the extended Yule–Walker equations by $\hat{b}_0^2$ gives the coefficients $[1, a_1, a_2]$. Dividing p_t (the least squares inverse developed in Section 9.1) by p_0 must also yield $[1, a_1, a_2]$. This is called the prediction error filter (PEF); here the PEF is of order 2.

The AR prediction model is $y_t = b_0 x_t - a_1 y_{t-1} - a_2 y_{t-2}$, so convolution with the PEF gives the time series of differences between the predicted and observed values, assumed to behave like white noise:

$$b_0 x_t = y_t + a_1 y_{t-1} + a_2 y_{t-2} = y_t * PEF$$

With only two coefficients, $PEF * y_t$ might not behave like white noise, but we can expect it to approach this condition as the order is increased. Regardless of order, the PEF acts as a whitening filter that optimally turns the time series y_t into white noise given the number of coefficients set by the chosen filter order.

9.5 Deconvolution Filters in Reflection Seismology

In seismic reflection data processing the PEF is known as a deconvolution filter, assumed to be an inverse filter to the wavelet in a linear filter model of the reflection seismogram; in such a model the seismogram is taken as the convolution of a linear filter l_t, also known as a wavelet, with a sequence of white-noise-like reflection coefficients. The reflection coefficients are represented by x_t, and t is the travel time between seismic source and receiver. The presence of additive noise is assumed, so that the model for the observed seismogram signal y_t is a linear filter relationship:

$$y_t = l_t * x_t + \text{noise}$$

This is the same model that underlies a linear filter description of ghost and water-bottom reverberations, as given in Section 8.6. As a result, we can view the wavelet l_t as a cascade of many filters, starting with the output of a seismic source (such as an explosion or a Vibroseis-sweep autocorrelation time series) and followed by filtering effects that occur along the travel path, including source and receiver ghosts, water-bottom and other reverberations (multiple reflections), and attenuation.

If the reflection coefficients do behave approximately like white noise, then the autocorrelation of the seismogram is proportional to the autocorrelation of the wavelet, within a scale factor, so that

$$y_t \star y_t \approx l_t \star l_t$$

A numerical demonstration of this is left as an exercise. Because the PEF depends only on the autocorrelation, it can be found using the autocorrelation of the seismogram. This assumption makes it is possible to find a wavelet inverse filter even if the wavelet itself is unknown. This is an example of blind deconvolution.

In developing the wavelet inverse filter, the presence of additive noise in the seismogram is taken into consideration by inflating the diagonal elements of the autocorrelation matrix with noise variance (false white noise), as discussed earlier. This will prevent the deconvolution filter (PEF) from amplifying the noise at frequencies where the signal amplitude may be low. Even if the assumption that the reflection coefficients are white-noise-like is not correct, the PEF acts as a whitening or spectral balancing filter. This is useful in adjusting the frequency content of the seismogram, which may have been relatively small at some frequencies. For example, attenuation may suppress high frequencies, low frequencies may not be effectively generated by the seismic source, and ghosts or multiple reflections may amplify or suppress some frequencies. In this way, the PEF acts as an inverse filter for effects that may not have been explicitly accounted for in prior processing steps.

9.6 Power Spectrum Estimate from the PEF

A discussion of the power spectrum and estimation methods using the DFT appears in the next chapter. Briefly, the power spectrum describes how the variance of a time series is distributed over frequency, and here we can see that, because the PEF tries to whiten the spectrum of y_t, its power transfer function tends to be reciprocal to the power spectrum of y_t. For a PEF of order 2, a power spectrum estimate should be proportional to

$$\frac{b_0^2}{|1 + a_1 Z + a_2 Z^2|^2}$$

where $Z = \exp(-i2\pi f)$. An order-2 PEF would not have enough coefficients to whiten the spectrum, in the usual case, so the order of the PEF might be increased until the residual time series $PEF * y_t$ behaves sufficiently like white noise. Testing it for white-noise-like behavior may involve examining a histogram of residuals (looking for a bell shape to test for Gaussian behavior), or examining its autocorrelation. Another criterion for selecting filter order is to set the order of the PEF to be large enough to predict over a time span corresponding to the longest-period oscillation in the time series. In practice many different PEF orders might be chosen, and the spectrum examined for all. Varying the PEF order may affect the precise location of spectral peaks, as shown in the next chapter; PEF orders larger than half the time series length are rarely used. A power spectrum estimate formed from the PEF carries a variety of names, including maximum entropy, autoregressive, parametric, and PEF. The whitening property of the PEF also makes it useful for conditioning a time series in order to improve the DFT and coherence spectrum estimates.

9.7 Vibroseis and Matched (Correlation) Filtering

Matched filtering is a method of inverse filtering for recovery of signals generated by a controlled source transmitting waves in a noisy environment. Originally developed for radar, it is now a standard method in sonar and reflection seismology. In reflection seismology it bears the name Vibroseis, following development and patenting of the name by Conoco in 1954. The term chirped radar or sonar is used in those applications.

The matched filter method shares the same goals as inverse filtering via least squares, in seeking an inverse filter that optimally recovers the signal and does not amplify the noise. Matched filters can be developed from least squares and maximum likelihood arguments in a way that is similar to earlier discussions. Because the transmitted signal design is a key element of the method, an additional advantage is reduced power requirements for the source. For Vibroseis, this means reduced damage to ground surface and structures when generating seismic waves (when compared to the use of to explosives), and, for sonar and radar, reduced electrical power requirements and lighter-weight equipment. This is especially important in the satellite radar systems used in radar altimetry to measure sea surface height and global mean sea level, as was illustrated in Chapter 1 and discussed further in Chapter 7.

Figure 9.1 A Vibroseis source (shaker) consists of a large vehicle with a hydraulic system to shake the ground with a prescribed sweep. The photograph shows the University of Texas Vibroseis source used in civil engineering applications to measure near-surface shear wave speeds and other properties related to site response during an earthquake. It was acquired with funding from the US National Science Foundation. The shaker is tri-axial (and known as TRex) and is able to shake along three axes (two horizontal directions, and the vertical direction). The majority of Vibroseis shakers are vertical only. During shaking the entire weight of the vehicle (about 30 metric tonnes) is used to hold down the central plate. The source signal can be a prescribed sweep such as that in Figure 9.2, or a continuous single-frequency sinusoid. Photograph courtesy of the University of Texas Department of Civil and Architectural Engineering.

The model for a Vibroseis seismogram signal y_t is a linear filtered version of the reflection coefficients x_t with added noise. The time series values are samples of the two-way time (surface to reflector at depth and return):

$$y_t = l_t * x_t + n_t$$

An example of a Vibroseis source (shaker) is shown in Figure 9.1. The shaker vibrates the ground to create an outgoing waveform l_t (known as a sweep) designed to have several important properties. First, l_t contains the range of frequencies of interest, set by the local geology, depth to target, seismic attenuation properties, and other considerations, such as the capabilities of the shaker's hydraulic and mechanical systems; outside the frequency band of interest, l_t is designed to have a near-zero output. Second, l_t is long in duration (many seconds) keeping the power requirements low. Total energy increases with duration, but the rate of energy transmission (power) is kept low. This also means that amplitudes are small so that linearity can be assumed. Linearity assures us that shaking at one location twice produces two seismograms that, when added together, will be identical to one from a shaker with twice the output. Third, l_t has an autocorrelation that is an approximate impulse. A correlation filter applied to the raw (uncorrelated) seismogram has the effect of compressing the long-duration Vibroseis sweep into an impulse-like autocorrelation, similar to the action of the least squares inverse filters discussed earlier.

After the seismogram is recorded, the sweep signal is correlated with it. The correlation filter applied to the seismogram is an MA filter whose coefficients are the time-reversed sweep signal, l_{-t}. The correlated Vibroseis seismogram signal is

$$l_t \star y_t = l_{-t} * y_t = (l_t \star l_t) * x_t + l_t \star n_t = l_{-t} * l_t * x_t + l_{-t} * n_t$$

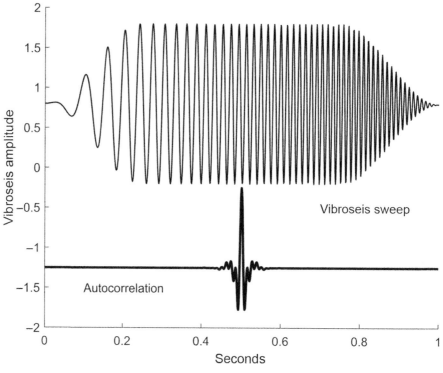

Figure 9.2 The Vibroseis sweep is a variable-frequency signal whose autocorrelation is impulse-like. Here a one-second sweep is used for illustration. In practice much longer sweeps of 10 or 15 seconds duration are used in exploration for petroleum. The impulse-like autocorrelation of the sweep is shown below it. The sweep signal contains oscillations increasing in frequency linearly in time from 10 to 100 Hz. When the sweep signal is correlated with the reflection seismogram, each occurrence of the sweep in the seismogram is collapsed to the impulse-like autocorrelation. Correlation is an MA filtering operating whose transfer function in Figure 9.4 passes the frequencies that were transmitted, thereby rejecting frequencies outside the 10–100 Hz range, which must be noise. The long-duration sweep means that the peak power is low, but the total transmitted energy is high. With low peak power, the Vibroseis source does little damage to the near-surface, so may be used in close proximity to structures and on paved roads, unlike other seismic sources such as dynamite. The sweep may be repeated several times at the same location to improve the SNR. It is also common for several Vibroseis shakers to be deployed in a spatial array to generate seismic waves in a synchronized shaking. A spatial array of Vibroseis sources generates fewer horizontally traveling surface waves, which are a nuisance in reflection seismology.

The first term is an approximation to x_t because l_t was designed in such a way that its autocorrelation $(l_t \star l_t)$ is an approximate impulse. Therefore, the correlation filter compresses the long-duration sweep, making it a pulse-compression filter. Figure 9.3 shows that the raw seismogram is not easily interpreted, but, after application of the correlation filter, distinct arrivals are associated with the reflection coefficients. Each reflection coefficient produces a copy of the impulse-like autocorrelation shown in Figure 9.2. In the above expression, the term $l_t \star n_t$ is the filtered noise. The noise has not been eliminated, but the correlation filter rejects noise at frequencies outside the range contained in the transmitted sweep l_t. Figure 9.4 shows the transfer function of the correlation filter corresponding to the sweep in Figure 9.2. The correlation filter rejects frequencies outside the range contained in the sweep because they were not generated by the Vibroseis source and must be noise.

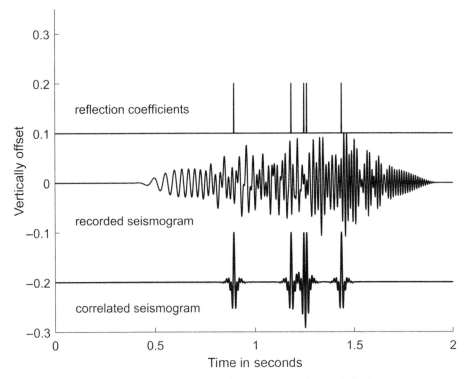

Figure 9.3 A simulated Vibroseis seismogram (center) is the correlation of the sweep signal with several reflection events. The reflected sweeps, which are of long duration, overlap each other. The correlated seismogram, at the bottom, resembles the reflection-event time series at the top because the correlation filter compresses every instance of the sweep into its impulse-like autocorrelation.

Correlation filtering (matched filtering) thus acts both as an inverse filter (compressing the sweep to an approximate impulse) and as a noise rejection filter.

9.8 Correlation Filtering in the Global Positioning System

A correlation filter can be used for precise arrival time measurements. The process is similar to a matched filter, but lacks the noise suppression feature found in Vibroseis where the correlation operation rejects frequencies outside the band of the transmitted signal. An important application is in the Global Positioning System (GPS) and similar satellite navigation systems. The GPS achieves one-way ranging by finding the time delay between signals transmitted from satellites and a receiver on the Earth's surface or in an aircraft or satellite below the GPS satellite altitude. The time delay is converted to an apparent distance (called the pseudo-range) by multiplying by the speed of light. The pseudo-range is then corrected for atmospheric and ionospheric conditions that cause the speed of light to vary.

Each satellite in the GPS constellation transmits a signal containing a pseudo-random binary code. The code is generated by a known algorithm, and is similar to white noise in that its autocorrelation

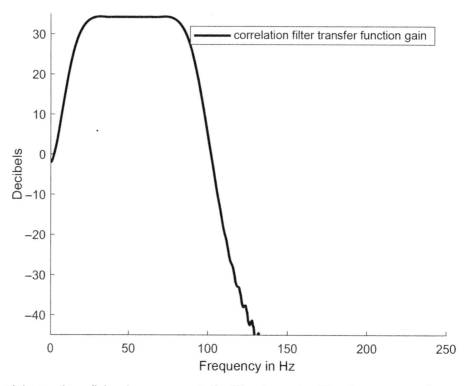

Figure 9.4 The correlation operation applied to seismograms generated by a Vibroseis source is an MA moving average filter whose pass band is that of the sweep signal. The transfer function here is computed from the squared modulus of the DFT of the sweep in Figure 9.2.

is nearly zero except at zero lag. Each satellite transmits a different code. The algorithm for creating these white-noise-like signals is in the ground-based receiver, where a copy of the code for each satellite is generated. Pseudo-random binary codes have a very broad bandwidth, so, unlike the Vibroseis system, GPS noise rejection is accomplished using a narrowband filter to select the carrier frequency upon which signals are encoded. A GPS receiver measures arrival time by correlating the locally generated copy of the code with the received signal. The arrival time occurs when the correlation is a maximum. The GPS signal also contains an almanac, with information on precise orbit locations for the entire GPS constellation. A schematic pseudo-random binary code and autocorrelation are shown in Figure 9.5. Additional noise suppression is achieved by increasing the length of the correlation time window. Like Vibroseis, the correlation filter acts as a pulse compression filter. Because the clock in a GPS receiver is of relatively poor quality, a complete position solution requires measuring arrival times from at least four satellites, three needed to determine location in three-dimensional space, and at least one more to account for clock error in the receiver.

Other applications of correlation filtering are found in seismology. One example is determination of the precise arrival times of seismic waves in seismic tomography, described in Chapter 7. Global-scale seismic tomography employs Earth models with thousands of voxels and uses thousands of seismograms. This requires an automated method to determine the arrival times of waves that have followed known ray paths passing through the Earth. The arrival times of waves are difficult to determine from actual seismograms. Figures 1.6 or 4.3 show that this is the case because arriving

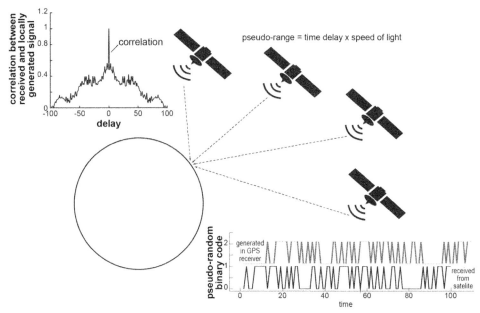

Figure 9.5 The Global Positioning System (GPS) determines locations using a correlation filter operating within a GPS receiver. The correlation filter determines the travel time from each satellite to the receiver, a quantity known as the time delay. The central portion of this figure depicts Earth as a circle, showing four GPS satellites transmitting to a point on the surface whose location is to be determined. The approximate length of each of the four dashed lines, called the pseudo-range, is the time delay multiplied by the speed of light. The pseudo-range must be corrected for variations in the speed of light due to atmospheric effects. The pseudo-range to three satellites would be sufficient to determine the location of the point, except that the receiver clock is not very accurate, so a fourth satellite is needed to compute a clock correction for the receiver. At lower right are two schematic time series, called pseudo-random binary codes. One of these is generated locally in the receiver; the other is that received from one of the satellites. Pseudo-random binary codes have autocorrelation properties similar to white noise, represented schematically in the upper left portion of the figure. This means that there will be a large correlation value between the local and received codes only when the locally generated code precisely aligns with the received code. Finding the delay that achieves a large correlation value determines delay time for that satellite.

waves emerge slowly. To provide a consistent and accurate measure of arrival time, a predicted waveform for a particular seismic event is correlated with the observed seismogram. The time of maximum correlation provides a consistent and objective measure of arrival time. Another application is in the field of ambient-seismic-noise seismology. Instead of using seismic waves generated by a source such as Vibroseis, ambient-noise seismology uses correlation to measure arrival time differences between seismometers at varying locations.

9.9 De-Blurring Filter Design

We now consider finding the inverse filter to the following MA filter:

$$y_t = x_t + x_{t-1} + x_{t-2} + x_{t-3} + x_{t-4} = [1, 1, 1, 1, 1] * x_t$$

The output y_t is a sum of five input time series values. As a practical example, suppose y_t represents samples along a spatial profile (now t is distance, not time). Then x_t represents the true values of the sampled quantity, and y_t the measured values. For example x_t might be pixels of an image along the profile, while y_t is the observed image taken by a camera that moved five pixels along the profile sampled by t while the camera shutter was open. In this case y_t is a linear-motion-blurred version of the true image x_t. In other cases, y_t might be a measurement of a quantity x_t obtained with a poor spatial resolution sensor. The goal is to find an inverse filter to remove the spatial averaging (blurring) effect.

From Chapter 5 we know that the exact inverse filter can be obtained by interchanging the roles of the MA and AR coefficients. Therefore the filter equation for the exact inverse is

$$y_t = x_t - y_{t-1} - y_{t-2} - y_{t-3} - y_{t-4}$$

The first 15 terms of the impulse response of this filter appear in Figure 9.6. They continue to repeat indefinitely, so the exact inverse is a metastable filter. Figure 9.7 shows that the transfer function of the blurring filter goes to zero at two frequencies, so that exact de-blurring will not be possible because those frequencies are completely absent in the measured output. The exact inverse transfer function, also shown in Figure 9.7, becomes infinite at those frequencies and is still large at nearby frequencies. Applying the exact inverse would amplify noise that might be present in the blurred image at spatial frequencies where the signal level is low, near the zero values of the transfer function.

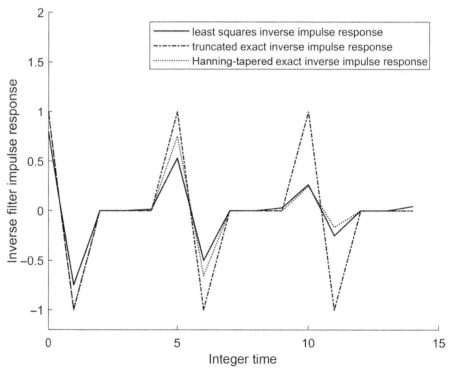

Figure 9.6 Impulse responses of inverse (de-blurring) filters showing the exact inverse, truncated to length 15, the least squares inverse filter coefficients of length 15, and the NB Hanning-tapered version of the exact impulse response. The Hanning-tapered version is very similar to the least squares result.

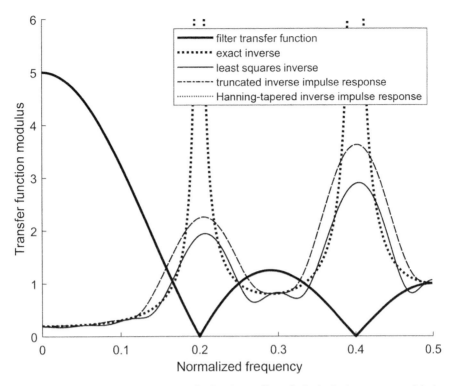

Figure 9.7 Transfer functions of the blurring filter, its exact inverse, the three inverse filters obtained using least squares, and the length-15 impulse response of the exact inverse. The transfer functions of the least squares and Hanning-tapered inverses are nearly identical and not distinguishable on this plot. The least squares inverse filter can be modified to account for the presence of noise by inflating the diagonal elements of the normal equation autocorrelation matrix, as discussed in the text and in the excercises.

An approximate inverse is needed to deal with these deficiencies in the exact inverse. Least squares is useful for this purpose, as well as the method of truncating the sampled impulse response described in the previous chapter.

A least squares MA inverse filter of length 15 was determined using the normal equations from Section 9.1. Its coefficients (equal to its impulse response) are plotted in Figure 9.6, and it can be seen that they are effectively a tapered (windowed) version of the exact-inverse impulse response series. This suggests another approach to finding an approximate inverse. We might simply find the exact-inverse filter, and apply a suitable one-sided taper (Hanning in this example) as used in the previous chapter. The resulting MA filter coefficients are indeed very similar to those of the least squares inverse filter, as can be seen in Figure 9.6. Figure 9.7 shows that the transfer functions of the least squares inverse and of the Hanning-tapered exact-inverse impulse response are nearly the same, and not easily distinguished in the plot. For reference we also include the transfer function of an inverse filter consisting of the first 15 values of the exact-inverse impulse response, without tapering. Its transfer function is similar, but the magnification at missing frequencies is considerably greater.

In practice, least squares provides additional flexibility in developing an inverse filter. As described in the first section of this chapter, inflation of the diagonal values of the normal equation matrix can be used as a design tool to reduce amplification of noise or to deal with unwanted artifacts

Figure 9.8   Illustration of de-blurring filter application and design. The upper left panel shows a portion of the original photograph in Figure 7.3, and to its right a blurred version after application of the filter $y_t = x_t + x_{t-1} + x_{t-2} + x_{t-3} + x_{t-4}$ in the horizontal direction. A least squares 15-term MA de-blurring filter has been applied in the horizontal direction to obtain the lower left image. The blurring has largely been removed, but there are filter artifacts (ghosts of image features) that are undesirable. The image ghosts to the right of the wheels are examples. There is no noise, so the artifacts arise because the transfer function of the de-blurring filter amplifies signals near the missing spatial frequencies shown in Figure 9.7. A modified least squares inverse filter is obtained by inflating the diagonal elements of the normal equation matrix by 10 percent. This is a damped least squares solution, also termed "adding false white noise" when used in seismic deconvolution applications. The lower right panel shows the de-blurred image using the modified filter. It has sharpened the blurred image nearly as well as the simple least squares filter, but has reduced the artifacts, although they have not been eliminated. Additional trial values of the inflation percentage might be tested to improve the de-blurred image further.

in the de-blurring process. This can be illustrated using the de-blurring of a photograph as an example. Examining the effects on the transfer function are left as an exercise. Figure 9.8 shows four versions of a portion of Figure 7.3. The top left panel is the original photograph. To its right is a blurred image, the result of applying along the horizontal direction the five-term blurring filter $y_t = x_t + x_{t-1} + x_{t-2} + x_{t-3} + x_{t-4}$. The lower left panel is the blurred image after filtering using the 15-term least squares inverse filter. The Hanning-tapered truncated impulse response filter would yield nearly identical results. Although the de-blurred image is reasonably sharp, it contains undesirable artifacts, evident as ghost images to the right of sharp edges. The bottom right panel shows the blurred image after application of a least squares inverse filter obtained by inflating the diagonal elements of the normal equation matrix by 10 percent. The ghost image artifacts have been reduced, with a slight reduction in sharpness of the de-blurred image. Further adjustment of the normal-equation matrix diagonal inflation might improve the de-blurred image.

9.10 Chapter Summary

This chapter has surveyed the use of least squares and related methods for designing linear digital filters and applications of correlation filtering. The main points are as follows.

- By expressing convolution as matrix multiplication, the inverse filter design problem takes the form of observation equations suitable for a least squares solution. The resulting normal equation matrix is banded and symmetric (Toeplitz) and contains the autocorrelation values.
- Least squares inverse filters depend only on the autocorrelation of the impulse response, and prediction filters depend only on the time series autocorrelation. As a consequence, deconvolution (inverse filtering) of reflection seismograms (as an important example) can be done blind, without knowledge of the wavelet.
- The task of predicting future values of a stochastic time series addressed by the Yule–Walker equations is not, at first glance, the same as finding an inverse filter. One discovers, however, that the extended Yule–Walker equations for the prediction error filter (PEF) have precisely the same form as the normal equations for the least squares inverse filter, showing the close connection between prediction and inverse filters.
- Besides its role as a deconvolution (inverse) filter, the PEF has other interpretations and uses. It is a whitening or spectral balancing filter, and it can be used as a power spectrum estimation tool. The use of the PEF in spectrum estimation is demonstrated in Chapter 10.
- Linear interpolation filters can be developed along the lines of the argument used to derive the Yule–Walker equations. Spatial interpolation is a common application, and many variants of this idea are in use, often described by terms such as kriging or geostatistics.
- The method of matched filtering can be developed from least squares principles, and is used in active source seismology (Vibroseis) and in chirped radar and sonar. In a matched filter the known transmitted waveform (sweep) is correlated with reflected signals both to compress the sweep to an approximate impulse, and to reject frequencies not transmitted, thereby rejecting noise.
- Correlation is also a powerful MA filtering method, used in GPS and other satellite navigation systems and for determining precise wave arrival times in seismic and other applications.

Exercises

9.1 **De-Blurring Filter.** The blurring filter MA coefficients are $b = [1, 1, 1, 1, 1]$ and the AR coefficient is $a = [1]$. The filter coefficients $[bi, ai]$ for the exact inverse to this filter are found by interchanging the MA and AR coefficients so they are $bi = [1]$ and $ai(1) = [1, 1, 1, 1, 1]$.

    A. Show that the exact inverse is metastable – its impulse response is oscillatory, never decaying to zero.

    B. As in the text, find a least squares MA inverse filter of length 15, and compare the transfer function as in the text with that of two other filters obtained when you inflate the diagonal elements of the Toeplitz autocorrelation matrix (by 10 and by 20 percent). Inflating the

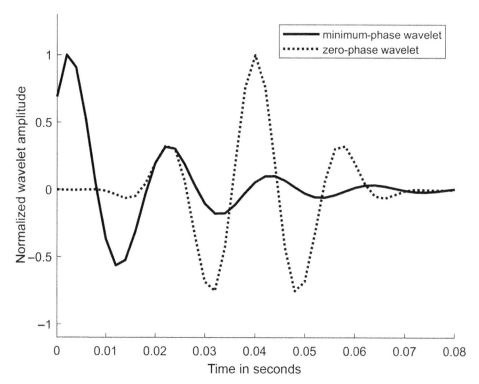

Figure 9.9 Zero-phase and minimum-phase wavelets. The minimum-phase wavelet is the first 41 values of the impulse response of an AR filter whose coefficients are $b = [1]$ and $a = [1.0000, -1.4562, 0.8000]$. The zero-phase wavelet is symmetric about its peak, which occurs at time zero, although it is plotted here with a delay to show it with the minimum-phase wavelet, which starts at time zero. The zero-phase wavelet is the central 41 values of the autocorrelation of a Vibroseis wavelet similar to that in Figure 9.2.

diagonal elements will cause the average gain to change at all frequencies, so normalize all inverse filter transfer functions to have the same gain as the exact inverse at zero frequency, and plot the power transfer functions (in dB) of the blurring filter, its exact inverse, and the three least squares approximate inverse filters (0, 10, 20 percent diagonal inflation) to show how inflating the diagonal elements of the autocorrelation matrix (adding false white noise) modifies the inverse filter and provides another tool in filter design.

9.2   **Shaping Filters.** Two seismic reflection surveys (surveys 1 and 2) are available for a given prospect, but each has a different wavelet, a minimum-phase for survey 1 (from a dynamite survey) and a zero-phase for survey 2 (done with Vibroseis). A seismic interpreter working with both data sets would like them to have the same wavelet, so a filter is needed to convert one wavelet into another. This filter is then applied to all seismograms of survey 1, so any given reflector will have the same appearance in both sets of seismic data. The zero-phase wavelet is preferred because its peak occurs at the time of the reflection event. Both are shown in Figure 9.9 and are 41 samples in length. The minimum-phase wavelet is the impulse response of a stable AR filter. The zero-phase wavelet is the central 41 values of the autocorrelation of a Vibroseis sweep similar to that in Figure 9.2.

A. Find a least squares shaping filter of length 50 to convert the minimum-phase wavelet into the zero-phase wavelet.

B. Plot the convolution of the shaping filter with the minimum-phase wavelet to show how well it works. You will need to pad zeros onto the front and back of the zero-phase wavelet when setting up the least squares observation equations. When comparing the results, there will also be a need to truncate the time series to have the same length, and to shift them in time.

9.3 Deconvolution Filters. It was stated in Section 9.5 that if a reflection seismogram is the convolution of a wavelet with white-noise-like reflection coefficients then the autocorrelation of the wavelet is approximately the same as that of the seismogram, thus allowing the wavelet inverse to be estimated from the seismogram. Using the minimum-phase wavelet from Exercise 9.2 and a set of reflection coefficients that you create using a random number generator for a time series of length 1000,

A. Compute the autocorrelation of the seismogram. This will result in 1999 terms from a seismogram of length 1000. Then select the 81 values centered on the zero-lag value, and scale them so that this zero-lag value has the same magnitude as that of the autocorrelation of the length-41 minimum-phase wavelet of the previous problem. Then plot together the autocorrelation of the minimum-phase wavelet (which is of length 81) and the 81 values from the seismogram autocorrelation to confirm that they are similar. This demonstrates that the autocorrelation of the wavelet may be reasonably estimated from the seismogram without knowing the wavelet itself.

B. Use values from the seismogram autocorrelation from part A to find an inverse (deconvolution) filter. Then convolve with the length-41 minimum-phase wavelet of problem 9.2 to show that it is a reasonably good inverse filter, resulting in an approximate impulse. This is a computational demonstration of blind deconvolution, as discussed in Section 9.5. The scaling applied to the seismogram autocorrelation in part A affects only the magnitude of the result, which is not important, because there is repeated rescaling of magnitudes during seismic-reflection data processing.

9.4 Stable Inverse Filter.

A. Find a least squares inverse filter (the PEF using the Yule–Walker equations) to find a stable inverse of length 40 to the Hector Mines source time function, first removing the leading and trailing zeros as in Exercise 8.4. Convolve the source time function with this stable inverse to show how effective it is as an inverse filter.

B. Find a minimum-phase length-40 version of the source time function by finding (via the PEF from the Yule–Walker equations) an inverse to the inverse filter you found in part A. Scale this and the source time function so that each has a maximum value of 1 and plot them together to show that they are similar. Repeat the calculation of the impulse response of the exact inverse of this minimum-phase version of the source time function, as in Chapter 8. In this case, because it is minimum-phase the impulse response does not grow exponentially.

10

Power and Coherence Spectra

The power spectrum describes how the variance of a time series is distributed over frequency. The variance (mean squared signal value per time sample) is a broadband statistic measuring power, while the power spectrum at a given frequency is a statistic measuring the power in a narrow frequency band. Similarly, the coherence spectrum describes how the broadband correlation coefficient between two time series varies with frequency. The DFT is the main tool for estimating both the power and coherence spectra and will be our main focus, but we also compare the DFT (periodogram) results with estimates made using the prediction error filter (PEF) developed in Chapter 9. Using examples we show that PEF estimates tend to be smooth, but the choice of PEF order introduces some variability in these estimates. Periodogram spectrum estimates tend to be erratic but can be tamed at the expense of diminished frequency resolution. We describe standard methods of assigning confidence intervals to periodogram spectrum estimates.

10.1 The DFT Periodogram

The simplest and most widely used DFT spectrum estimate is called a periodogram, a term introduced in the late nineteenth century. The periodogram is proportional to the squared modulus of DFT values and is so commonly associated with power spectrum calculations that its connection with the DFT often goes unmentioned. To understand it, we start with the Parseval theorem relating the sum of squared time domain values x_t to the sum of the squared moduli X_m in the corresponding DFT.

With the DFT normalization adopted in this book, the Parseval theorem states that the DFT is a linear transformation that preserves length, so that

$$\frac{1}{N} \sum_{t=0}^{N-1} \|x_t\|^2 = \frac{1}{N^2} \sum_{m=0}^{N-1} \|X_m\|^2$$

The mean of x_t is set to zero, so the left-hand side is the time series variance. The Parseval theorem then gives this variance as the sum of contributions from every Fourier frequency on the right-hand side. Each term in this sum is the variance within a frequency band of nominal width $\Delta f = 1/(N \Delta t)$, centered at every Fourier frequency. The frequency domain sampling of a periodogram estimate is the transfer function of the boxcar window in Figure 6.2. The right-hand side of (10.1) is, however, just a raw periodogram estimate. Describing it as an estimate is noteworthy because a periodogram typically has erratic features making it a wild or untamed estimate. These erratic features, small peaks and troughs, do not usually imply sinusoidal variations at any particular frequency. Prior to computing the spectrum, one should remove from the time series any variations that are well understood, such as seasonal-frequency sinusoids and other variations that cannot be well represented using Fourier frequency sinusoids, such as linear, exponential, or other slowly varying trends. Using least squares to fit and subtract such features is the standard method to obtain a residual series suitable for spectral analysis. One goal of spectral analysis may then be to search for unknown sinusoidal variations. In a power spectrum, pure sinusoids are termed line spectra. Therefore one must be able to assess the significance of a peak in the spectrum. This is done by establishing confidence intervals, to be discussed shortly.

For real-valued time series it is only necessary to give the periodogram at positive frequencies, because negative frequencies are complex conjugates with the same squared modulus. Thus, spectrum values are multiplied by two to account for the omission of negative frequencies, excluding the Nyquist which only appears once in the DFT. The periodogram is then

$$S_m = \frac{2}{N^2}[\|X_1\|^2, \ldots, \|X_{(N/2)-1}\|^2, \tfrac{1}{2}\|X_{(N/2)}\|^2]$$

In this form the periodogram values have units of variance, but it is more common to use units of power spectral density (PSD) which are variance per frequency. Then the sum of all the PSD values multiplied by the frequency bandwidth $\Delta f = 1/(N\Delta t)$ (the spacing between DFT values) is the variance. The advantage is that PSD units are independent of time series length, allowing spectra from varied data (different record lengths or sampling frequencies), to be compared. To convert to PSD units, write the Parseval theorem (10.1) as

$$\frac{1}{N}\sum_{t=0}^{N-1}\|x_t\|^2 = \frac{1}{N\Delta t}\frac{\Delta t}{N}\sum_{m=0}^{N-1}\|X_m\|^2$$

The left-hand side has the units of variance, and the right-hand side is a sum of PSD values multiplied by bandwidth of each, $1/(N\Delta t)$. For frequencies $f = m/(N\Delta t)$, the PSD periodogram estimates are

$$P_m = \frac{2\Delta t}{N}[\|X_1\|^2, \ldots, \|X_{(N/2)-1}\|^2, \tfrac{1}{2}\|X_{(N/2)}\|^2]$$

It can be seen that at the Nyquist frequency $m = N/2$ the value is not doubled, with the result that

$$\frac{1}{N}\sum_{t=0}^{N-1}\|x_t\|^2 = \sum_{m=1}^{N/2}P_m\frac{1}{N\Delta t}$$

The PSD units are variance per Hz if time is measured in seconds. The square roots of the PSD values give the amplitude spectrum. With the same units as the original time series, the amplitude spectrum is often plotted instead of PSD. The bandwidth of each amplitude spectral value is the same as the PSD, but it is common for amplitude spectrum units to be given as amplitude per square root Hz, even though square root Hz does not make much sense.

10.2 Periodogram of White Noise

An interesting case that illustrates the erratic nature of the DFT periodogram arises when it is computed for a time series of zero-mean independent random numbers. The true power spectrum is constant at all frequencies, because the autocorrelation is, in the limit of large numbers of data, just equal to an impulse at zero lag and zero at all other lags. Because its spectrum is constant, a time series of random numbers is called white noise (analogous to white light). The variability of a white noise periodogram therefore serves as a null test, to establish confidence intervals so as to be able to judge when a spectral peak is significant in a periodogram computed from a finite-length time series. This is useful when searching for new line spectra. An example of a Gaussian white noise time series is shown in Figure 10.1.

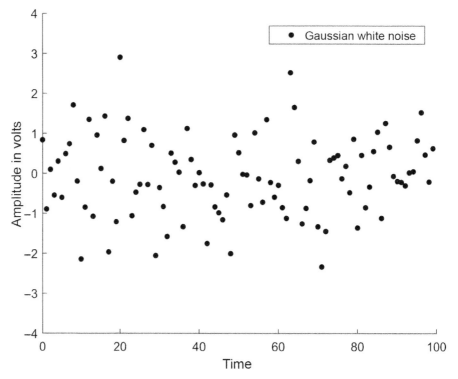

Figure 10.1 A Gaussian white noise time series generated by a pseudo-random number generator. Each time series value is independent of all others. The power spectral density of white noise is a constant at all frequencies. The periodogram of white noise is useful in illustrating the erratic nature of periodogram estimates and the trade-off between improved stability of estimates and reduced frequency resolution of the power spectrum. In addition, the white noise periodogram provides a null case to assess the significance of peaks in the power spectrum. This leads to confidence intervals that can be used when searching for periodic or near-periodic behavior in a time series.

For white noise, each raw periodogram estimate (proportional to the sum of the squares of the real and imaginary parts of X_m) behaves like a chi-squared random variable with two degrees of freedom. (Chi-squared variables are described in Appendix C.) The chi-squared variability associated with this null case (where the true spectrum is known to be white) provides confidence intervals as will be discussed shortly. Figure 10.2 shows the erratic nature of the periodogram PSD estimates when plotted on a linear vertical scale. Figure 10.3 shows the same PSD plotted on a decibel scale, as is more common and generally more useful. The dB scale is better at displaying the often very large range of numerical values of periodogram estimates, and it allows confidence intervals to be expressed in decibels. If a linear vertical scale were used, confidence intervals would be different at each frequency because they depend on the spectrum estimate, as will be discussed shortly.

One might suppose that the erratic behavior of the periodogram estimate would improve with more data, but this is not the case. For example, if the length of a time series is doubled, the estimates are just as wild, each behaving as a chi-squared random variable with two degrees of freedom. Now, however, there are twice as many estimates in the Nyquist band, providing an opportunity to average values at adjacent frequencies. Averaging two adjacent raw periodogram values yields a chi-squared random variable with four degrees of freedom, with greatly reduced erratic behavior. Equivalently (in a statistical sense), one can divide the series in two, compute a separate periodogram from each half and average them. Either approach to taming the wild periodogram reduces the

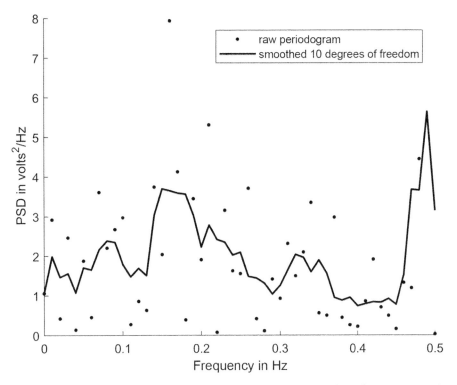

Figure 10.2 Raw and smoothed periodograms of the white noise series in Figure 10.1. The vertical scale is linear. By averaging periodogram values over five adjacent frequencies, the estimates are less variable. The random behavior of periodogram estimates is described by the chi-squared probability density function given in Appendix C. Raw periodogram estimates behave like chi-squared random variables with two degrees of freedom, while averaging five of them gives an estimate that behaves like a chi-squared value with 10 degrees of freedom. Averaging in frequency leads to reduced scatter in the estimate at the expense of poorer resolution in frequency.

frequency domain resolution in order to stabilize the estimates. This trade-off is a central element of spectral analysis, and may be of concern in some applications, for example, when searching for unknown line spectra. Averaging in frequency is justified if the true spectrum is known to be locally constant, though rarely does one know this in detail. Either averaging periodogram estimates in frequency or averaging periodograms computed from several portions of a longer time series requires an assumption that the time series is stationary. These considerations make spectrum analysis something of an art that benefits from the confirmation of stationarity, and may involve an iterative process to judge how much smoothing in the frequency domain is justifiable.

We can illustrate the determination of confidence intervals useful in judging whether a peak in the periodogram spectrum is significant. This question might arise when searching for unknown periodic behavior, for example in a climate time series. In the example white noise time series here (Figure 10.1) and in the next section for the case of colored noise, the digital sample interval is 1 second, the time series length is 100, and the spacing between Fourier frequencies is 0.01 Hz. In both cases a zero-mean unit-variance Gaussian white noise series was obtained using a computer random number generator.

Suppose that five raw periodogram estimates have been averaged to obtain 10 degrees of freedom. Chi-squared tables (Appendix C) show that in the case of 10 degrees of freedom a chi-squared variable falls within the interval $[3.247, 20.483]$ 95 percent of the time. If $\hat{S}$ is the smoothed

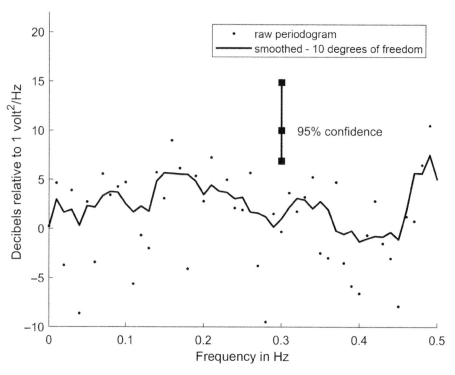

Figure 10.3 Periodogram power spectral density as in Figure 10.2, but here on a decibel scale, including confidence intervals for the smoothed estimates. Using a decibel rather than a linear scale, the confidence interval is the same for all estimates. When the spectrum is understood as an estimate from a finite set of data that might be improved with a longer time series, the confidence interval is interpreted as the interval in which the true spectrum lies with 95 percent confidence.

periodogram estimate at some frequency and S is the true value of the spectrum at this frequency then the usual procedure is to assume $10\frac{\hat{S}}{S}$ is chi-squared distributed with 10 degrees of freedom. With this assumption the properties of chi-squared variables (Figure C.4) are such that

$$3.247 < 10\frac{\hat{S}}{S} < 20.483$$

with 95 percent confidence. The factor 10 goes with the definition of a standard chi-squared variable as the sum of the squares of 10 squared unit-variance Gaussian random variables. The inequality is then rearranged to express the true value in an interval surrounding the estimate $\hat{S}$:

$$0.49\hat{S} < S < 3.08\hat{S}$$

Chi-squared random variables have the property that the confidence interval is proportional to the estimate. This means the confidence interval varies across frequency because the estimates are likewise variable. This is inconvenient to display, and it is usually easier to express the confidence interval using a decibel scale, giving the spectrum in dB, and the factors 0.49 as about -3 dB and 3.08 as about $+5$ dB surrounding $\hat{S}$. Confidence intervals are shown in Figure 10.3. Another advantage of the decibel scale is its value in compactly representing the large range of numerical values typically obtained in periodogram estimates. In this case, if we found that the spectral peak was not more than 5 dB above the level at the surrounding frequencies, we would say that the peak is not significant at the 95 percent level, because the true value could be (with 95 percent confidence) about the same as the spectrum at the surrounding frequencies.

10.3 Comparing Power Spectrum Estimation Methods

In general terms a spectrum may include two types of behavior. One type, found in some geophysical time series, is associated with purely sinusoidal variation. As examples, climate series show daily or annual changes, and sea level shows tidal and other changes due to periodic forcing. Purely sinusoidal variations create in a power spectrum discrete line spectra, and are distinct from a spectral continuum. Power spectrum analysis is applicable to both line and continuous spectra, but it is important to recognize the presence of line spectra when computing and interpreting spectrum estimates. Coherence spectrum estimates can also be biased by the presence of line spectra. If the underlying physics suggests purely sinusoidal variations, these should usually be removed, typically by fitting sinusoids at the corresponding frequencies via least squares. After their removal, the residual time series should have a continuous spectrum suitable for analysis by DFT and other methods. Removing line spectra is one example of conditioning used to improve DFT spectral estimates. Additional conditioning may include whitening to balance the spectrum, and the PEF could be used as a whitening filter for this purpose. In addition one should remove linear trends and other low-frequency variations (with oscillations at periods near to or greater than the time series length) using least squares.

A flaw of the periodogram is that it tends to blur sharp spectral peaks and to fill in spectral troughs (low points). The blurring of peaks and filling of troughs can be described as a leakage of spectral power across frequencies. Section 6.4 describes its cause as the frequency domain convolution

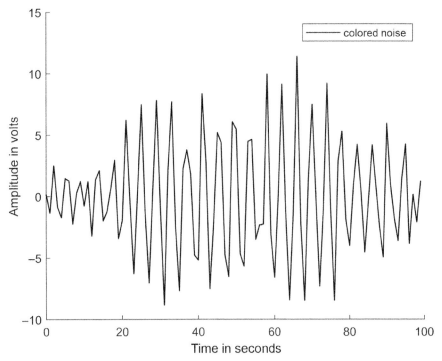

Figure 10.4 Colored Gaussian noise, obtained by applying the ARMA filter described in the text to a Gaussian white noise time series. Because the spectrum of Gaussian white noise is known to be constant, the spectrum of the colored noise is just the power transfer function of the applied filter. The digital sample interval is 1 second, so the Nyquist frequency is 0.5 Hz, and the nominal DFT frequency resolution is the reciprocal of the record length, equal to 0.01 Hz.

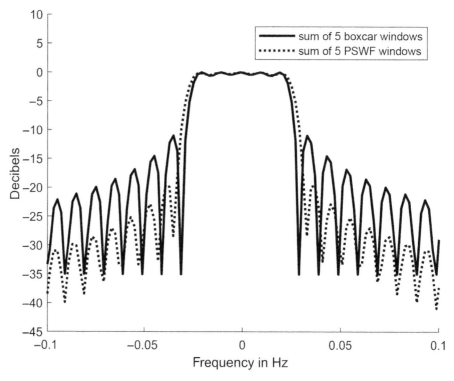

Figure 10.5 Frequency domain windows for spectrum estimates with 10 degrees of freedom, from the multi-taper method (MTM) (five PSWF tapers, as shown in Chapter 6) and for an average of five adjacent periodogram estimates. The MTM window is flat in the pass band and has much lower side lobes relative to the simple periodogram average. The periodogram window, the sum of five adjacent windows, has ripples in the pass band and side lobes that are about 10 dB higher than those of the MTM estimate, which is the sum of five PSWF eigenspectra windows.

between the spectrum and a window transfer function. The severity of spectral leakage depends on the sharpness of the troughs and peaks and on the record length, with its associated frequency resolution proportional to reciprocal record length. Chapter 6 discusses other window choices, including the PSWF windows used in multi-taper method (MTM) spectrum estimates. The reciprocal of the power transfer function of the PEF also provides an autoregressive spectral estimate, as discussed in Chapter 9, but this requires choosing the order of the AR filter. With these various choices available for power spectrum analysis, it is useful to show how the various methods perform using an example time series where the true spectrum is known exactly. This is similar to the previous case (the periodogram of white noise), but now the time series is colored Gaussian noise, that is, Gaussian white noise filtered by an ARMA filter. The coefficients of the filter in this example are

$$a = [1.0000, -0.0017, 0.9490]$$
$$b = [1.0407, -1.4715, 1.0407]$$

The associated power transfer function of this filter (the true spectrum) is shown in Figures 10.4, 10.5, 10.6, and 10.7.

The true spectrum has a peak at 0.25 Hz and a trough (zero value of the spectrum) at 0.125 Hz. Four spectrum estimates are computed. The first is a simple periodogram (boxcar window) averaged over five DFT values to obtain 10 degrees of freedom. The second is an MTM estimate, using the

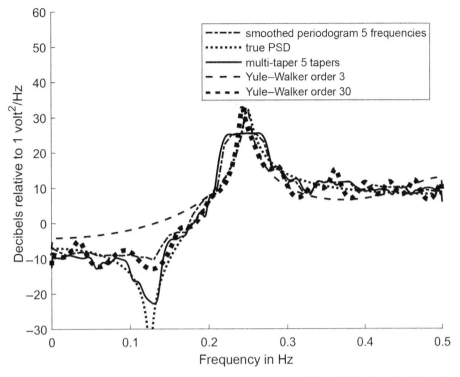

Figure 10.6 The true spectrum of colored Gaussian noise and estimates using the periodogram, MTM, and PEF methods. The PEF estimates (the PEF is obtained using the Yule–Walker equations) are very effective at identifying the spectral peak, but not the trough. The MTM estimate is effective at identifying the trough, but averages considerably in frequency, so is not useful in identifying the precise location of the peak. Both the MTM and periodogram estimates reflect averages over a frequency bandwidth of about 0.05 Hz, and so are not suitable for identifying the location of the spectral peak.

five discrete prolate-spheroidal windows in Figure 6.3 to obtain five eigenspectra, which are then added together giving an equivalent 10 degrees of freedom. The frequency domain windows of the simple periodogram and MTM estimates are shown in Figure 10.5. The MTM window is the same as in Figure 6.4, but is plotted here on a decibel scale. The MTM and periodogram methods provide similar averaging in frequency, but the MTM window has much lower side lobes, and lacks the ripples in the pass band associated with the sum of five adjacent boxcar windows. The central boxcar window is shown in Figure 6.2. The third spectrum-estimation method uses the PEF (from the Yule–Walker equations), with PEF orders of 3 and 30. The time series length in this case is 100, so a PEF of order 30 is then 30 percent of the time series length.

   Figure 10.6 shows all the estimates and the true spectrum computed from the ARMA filter transfer function. Figures 10.7 and 10.8 show details of the trough and peak regions. The PEF for order 3 does not get the peak location quite correct, but for order 30 it is closer to correct. The MTM and periodogram are very similar at most frequencies; they severely smear the spectral peak because the time series is quite short, and averaging over five adjacent frequencies (or five eigenspectra) further degrades the resolution of the peak. Only the AR (PEF) method provides reasonable estimates of the peak. However, only the MTM method is able to correctly detect the spectral trough near 0.125 Hz. The other estimates fill in the trough due to the leakage of variance from adjacent frequencies.

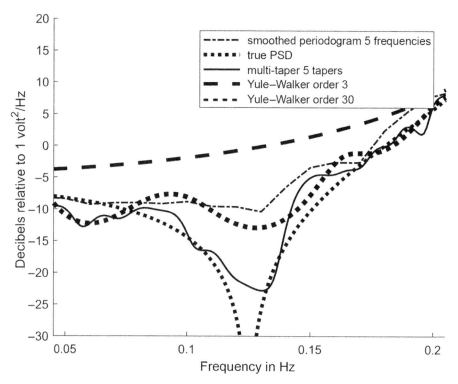

Figure 10.7 Details of the spectrum estimates from Figure 10.6 around the trough.

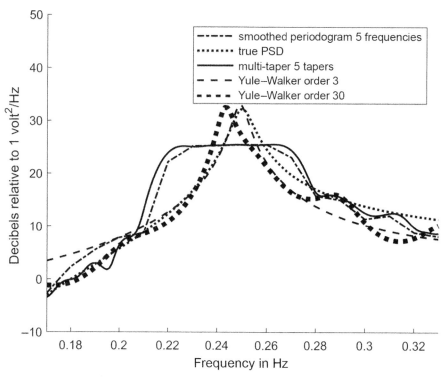

Figure 10.8 Details of the spectrum estimates from Figure 10.6 near the spectral peak.

10.4 Correlation and Coherence

A standard problem in time series analysis is to determine whether and to what degree two time series are related or partially related to each other. The correlation coefficient (the Pearson correlation coefficient) and its square are commonly used to measure this. In the general case, two series may be only partially related, so that each contains a related component but either or both contain noise (or unrelated variations), which diminishes their correlation. Appendix C shows examples where the numerical value of the correlation coefficient can be related to the signal to noise ratio (SNR) in certain cases.

Suppose that time series x_t and y_t are related by $y_t = lx_t + n_t$, where l is a scalar and n_t is another series unrelated to x and y. The correlation coefficient r_{xy} is calculated after subtracting the mean from each series. One computes the cross correlation at zero lag and then normalizes by the standard deviations of each series. This calculation may also be understood as forming the dot or inner product of the time series x and y, each of which has been normalized to unit length by dividing by their standard deviations. As a dot product between two unit-length vectors, the value will be less than unity in magnitude and may be positive or negative:

$$r_{xy} = \frac{\sum_{t=0}^{N-1} x_t y_t}{\left(\sum_{t=0}^{N-1} x_t^2 \sum_{t=0}^{N-1} y_t^2\right)^{1/2}}$$

Negative values of the correlation coefficient r_{xy} imply that the scalar l is negative.

The correlation coefficient may be biased in various ways, either appearing large when the time series are unrelated or small when they are not. For example, if both time series contain linear trends, the correlation coefficient may be large even though the trend is the only common feature. Similarly, if one time series has been shifted relative to another, the correlation coefficient may be small even though the two series are identical except for the shift; in the example of the GPS system, when time series behave like white noise, the correlation coefficient magnitude goes nearly to zero. It is important therefore to first remove linear trends and known time shifts. In addition, the correlation coefficient may also be inflated when both time series contain a common sinusoidal constituent. If this is known in advance, it is best to remove this sinusoidal term from each series.

A standard assumption about the time series x_t and y_t is that each has a white spectrum, but this is rarely true. Interpretation of the numerical value of the correlation coefficient, and the confidence intervals for the true correlation coefficient when the spectrum is not white must be determined by Monte Carlo methods.

A more general approach to measuring correlation is to assume that the two time series are related via a linear filter,

$$y_t = l_t * x_t + n_t$$

and to use $R_{xy}(f)$, the spectrum of coherence, to estimate the correlation coefficient (or its square) as a function of frequency. The coherence spectrum $R_{xy}(f)$ is a complex-valued normalized cross spectrum, usually presented in polar form as the squared modulus (squared coherence) and phase.

When the time series are incoherent, the phase is typically random and so appears uniformly distributed over the range $[-\pi, \pi]$.

Estimates of the coherence and phase spectra are computed from the DFT of each series. One must decide on a compromise between the statistical confidence or stability of the coherence estimate, and the need for good resolution in frequency. Coherence estimates are found by averaging across several DFT values, so all the considerations important in periodogram spectrum estimation are relevant. Coherence estimation is preceded by conditioning the time series, including the removal of means and trends, line spectra, and possibly additional pre-whitening, using the same pre-whitening filter for both time series. Without pre-conditioning and pre-whitening, coherence estimates may be biased (inflated or deflated relative to their true value). Known time shifts (such as propagation delays) should be removed because these can also bias estimates, making them too small. Removing time shifts is a "pre-whitening" of the phase. Time domain tapering windows can be used but multi-taper PSWF windows are preferred because of their superior frequency resolution and small leakage across frequencies.

To form estimates of coherence and phase spectra between x_t and y_t, one computes an average cross spectrum either by averaging over frequency (with a single conventional window) or averaging multi-taper-method eigencross spectra $X_f Y_f^*$. Like power spectrum estimation, increased averaging reduces scatter and shrinks the confidence interval at the expense of reduced frequency resolution. For a simple (single conventional window) cross spectrum estimate near $f = m/N$, an average cross spectrum over five adjacent Fourier frequencies would be

$$S_{xy}(m) = \sum_{k=-2}^{k=2} X_{m+k} Y_{m+k}^*$$

The phase spectrum at this frequency is taken from the phase of the complex-valued cross spectrum. The squared coherence is the squared modulus of the cross spectrum normalized by the product of the periodogram power spectra $S_x(m)$ and $S_y(m)$, averaged over the same five frequencies. The coherence squared is

$$R_{xy}^2(m) = \frac{\|S_{xy}(m)\|^2}{S_x(m) S_y(m)}$$

The number of degrees of freedom for the squared coherence is twice the number of bandwidths, or twice the number of PSWF windows. Published charts allow the interpretation of the significance of the squared coherence, given the number of degrees of freedom. Alternatively, confidence intervals can be established by Monte Carlo experiments.

10.5 Coherence of Sea Level Variations

An example application of the coherence spectrum is in evaluating the relationship between sea level variations at different locations. The sea level time series in Figure 1.3 provide good examples for this; however, we consider here the coherence between the pair of time series in Figure 10.9, showing measurements of the monthly average sea level change at Los Angeles, California, and Victoria, British Columbia, over a 29-year period. Visual inspection of the series suggests they may be correlated at low frequencies, but not at high. This would be the case if the high-frequency (short-period) sea level change is forced by local weather conditions, while longer period changes have a

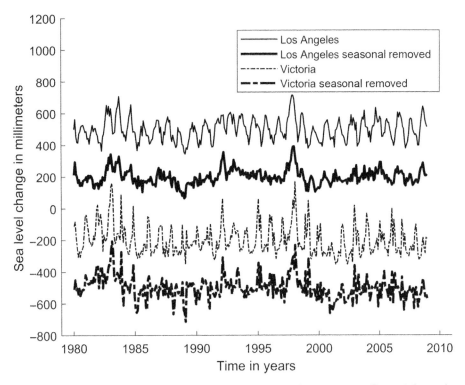

Figure 10.9 Twenty-nine years of monthly average sea level at Los Angeles and Victoria, two of the time series in Figure 1.3. Seasonal variations have been removed using the DFT by setting to zero coefficients at 1, 2, 3, 4, 5, and 6 cycles per year, as described in Chapter 4.

common origin in Pacific-Ocean-wide climate signals, such as El Nino or others. In fact this sort of analysis has been used to identify not only the ENSO (El Nino Southern Oscillation) variations that are Pacific-wide, but also longer decadal-period variations in the Pacific.

Here we compute the spectrum of coherence using the multi-taper method, with five tapers. The result is shown in Figure 10.10. It confirms that the squared coherence is about 0.5 or greater below about two cycles per year, and is effectively zero above this. This example calculation confirms what is evident visually.

10.6 Searching for Milankovitch Periods

It is generally agreed that one of the great geophysical discoveries of the twentieth century was recognizing that Earth's crustal deformation is largely described by global plate tectonics. Yet an equally important geophysical discovery was that major climate changes, including the ice ages, tend to occur at regular intervals. An important example is the climate variations at periods associated with Earth's orbital changes. These alter the amount of sunlight falling on the northern and southern hemispheres in summer and winter. The underlying theory was developed by Serbian geophysicist Milutin Milankovitch and was published in a series of articles and books, starting in 1920. Milankovitch predicted orbital-related climate forcing at periods of about 100-, 41-, and 26-thousand years. Although the theory was well grounded in physics, it was not widely accepted in

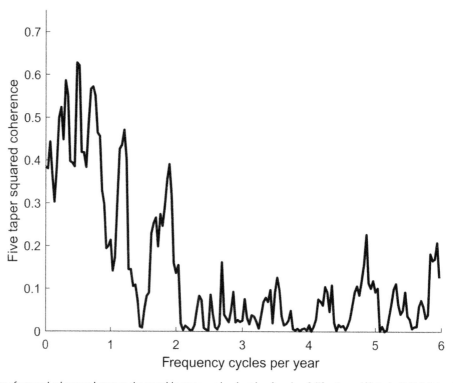

Figure 10.10 Spectrum of squared coherence between the monthly mean sea levels at Los Angeles, California, and Victoria, British Columbia, showing a coherence that is high below 2 cycles per year (cpy), and near zero at higher frequencies. This can be interpreted as being due to a common low-frequency source of sea level change in the Pacific basin, with the higher-frequency variations being controlled by local effects at the two sites. Local effects at high frequencies are not coherent because of the large distance between Victoria and Los Angeles. Even though true coherence is likely to be zero above 2 cpy, several spurious peaks are present. Such spurious peaks are typical of coherence spectrum estimates. As in power spectral estimation, coherence spectrum estimates require trading off the stability of estimates (to suppress spurious peaks) against the need for good frequency resolution.

his lifetime. About two decades after his death in 1958, the first geophysical evidence for periodic climate variations became available from deep-sea sedimentary cores. In the deep oceans, a slow but steady rate of sedimentary deposition allows relatively precise dating, while oxygen isotope and other data provide evidence of climate changes in the cores. The power spectral analysis of the time series derived from the cores then provided the first evidence supporting the Milankovitch theory.

In the decades since then, there has been growing interest and effort in gathering time series of past climate changes from various sources. Important detailed records have been obtained from Antarctic ice cores. An example is shown in Figure 1.5, the Lake Vostok core, extending to over 400 thousand years into the past. Here we examine the Vostok time series using power spectral analysis to illustrate applications of the methods surveyed earlier in this chapter. There are several goals in such an analysis. One is to confirm the presence of variations at the periods predicted by Milankovitch. By itself this would support the idea that a linear filter model is useful in describing Earth's climate system. That is, it would show that a single forcing frequency produces a single frequency variation in the climate. Even without spectral analysis, one can see from Figure 1.5 that a 100-thousand-year period is present in the time series, and is the dominant period of oscillation between glacial and inter-glacial episodes. The presence or absence of other Milankovitch periods is less obvious from visual inspection. If spectral peaks are not precisely at Milankovitch periods, the linear filter model may be questioned. If Milankovitch periods are found in the power spectra, it is important to measure their amplitudes relative to the magnitude of forcing. That is, it is important to measure the transfer function of the climate system (considered as a linear filter) at different periods. It is recognized that even though the 100-thousand-year forcing is relatively weak in the Milankovitch theory, it results in the largest climate variation. This implies a large transfer function near this period, which requires an interpretation in terms of climate physics. Both the precise frequencies of the power spectral peaks and the interpretation of the climate system transfer function are topics of continuing research.

Here we use the Lake Vostok time series to illustrate the use of the periodogram and autoregressive spectral methods in a search for peaks in the power spectrum near Milankovitch periods. This will also illustrate the distinctive properties associated with each method. The multi-taper method, examined earlier in this chapter, provides an averaged spectrum and is not well suited to finding an isolated spectral peak. This is evident, for example, in Figure 10.8. The multi-taper method is not employed here for that reason. For the two AR estimates, the PEF order is chosen to be 1000 and 1500. This provides PEF filter lengths of 100- and 150-thousand years, which are suitable for

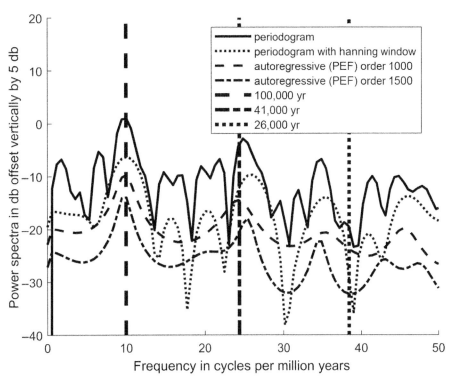

Figure 10.11 Power spectra of the Lake Vostok proxy-temperature time series in Figure 1.5. All estimates are plotted at the same frequency resolution. Each of the four estimates is offset vertically by integer multiples of 5 dB for clarity. For the periodogram estimates, the time series are padded with zeros to obtain the same frequency resolution as in the two AR estimates. All four estimates show evidence of peaks near 100- and 41-thousand-year periods, but none at 26-thousand years. The effect of the Hanning window on the periodogram estimates is to slightly broaden the 100-thousand-year spectral peak, and to reduce spectral leakage from nearby frequencies. Both would be expected from the broader central lobe and lower side lobes of the Hanning window transfer function (Figure 6.2) relative to the boxcar transfer function. The narrower peaks in the two AR spectra show that the AR method is better at resolving the precise frequency of a spectral peak. However, there is some variability of the peak frequency among the four estimates.

making predictions over the times between ice ages. Figure 10.11 shows strong peaks from all four estimates at the 100-thousand-year period. This is no surprise, as this time scale of variation is evident from visual inspection of the time series in Figure 1.5. All estimates are plotted at the same frequency resolution. For the two periodogram estimates, the frequency resolution is adjusted by padding the time series with zeros. All four estimates show evidence of a peak near the 41-thousand-year period, but none suggests one near 26-thousand years. The effect of the Hanning window on the periodogram estimates is to slightly broaden the 100-thousand-year spectral peak, while at the same time causing less leakage into adjacent frequencies. The narrower peaks in the two AR spectra imply a slightly better ability to resolve the precise frequency of each peak. There is some variability of peak location depending on the particular estimation method.

10.7 Chapter Summary

Spectrum analysis is concerned with describing how time series variance or correlation with another time series may vary with frequency. The main points are as follows.

- The standard power spectrum estimate is based on the periodogram and is proportional to the squared modulus of the time series DFT.
- The periodogram of white noise is useful as an example to understand why averaging is needed to stabilize power spectrum estimates, but this comes at the expense of diminished frequency resolution.
- The variability of power spectrum estimates for the white noise case provides a standard way to assign confidence intervals for spectrum estimates using properties of chi-squared random numbers.
- Power spectra are most commonly plotted on a vertical scale in decibels, both because the range of values is often large and because confidence intervals are most easily presented as a decibel range.
- There are a number of choices in forming periodogram estimates, so the preferred method depends on the task at hand. For example, seeking the precise frequency of a spectral peak in the search for Milankovitch periods requires a method with high resolution in frequency, so the multi-taper method (MTM) would not be suitable.
- Two time series may be correlated, but not at all frequencies. The spectrum of coherence is useful for discovering correlations in such cases. Like the power spectrum, coherence estimates can be erratic and require confidence intervals, which can be determined by Monte Carlo experiments using simulated data. Coherence estimates are tamed by averaging over frequency, at the cost of poorer frequency resolution.

Exercises

10.1   **Milankovitch Periods.** Read about the discovery of the Milankovitch cycles in global climate records, starting with analysis of deep-sea sediments in the 1970s, then proceed with an analysis of the Lake Vostok ice core proxy temperature record. First plot the time series noting the regular intervals of ice ages, and the end of the last ice age about 20,000 years ago. Its time values are measured in years before the present.

Compute the power spectral density using other choices for PEF order besides those used in this chapter: try choices of the values used in the chapter (1000 and 1500) and one-half those values. Find the value of the peak in the spectrum to determine how it varies with PEF order.

10.2 **Coherence of Sea Level.** Use the Pacific sea level data set, consisting of 29-year monthly samples of selected mid- and eastern Pacific tide-gauge records from the Permanent Service for Mean Sea Level. Examine the coherence between mid-ocean stations Midland and Honolulu, and between other eastern Pacific stations. Precondition all time series by removing the mean from each, and removing the seasonal component (sinusoids of $1, 2, 3, 4, 5, 6$ cycles per year). Because the seasonal variations are typically a large part of the signal, the seasonal signals would bias coherence estimates. Remove them using either least squares or the DFT. Compute the average squared coherence at frequencies below one cycle per year for pairs of time series. Determine whether nearby stations are more coherent at higher frequencies relative to the example involving Victoria and Los Angeles.

Appendix A Matrices and Vectors

The purpose here is to review basic properties and uses of matrices and vectors, which are essential tools in time series and data analysis. Two main uses of vectors and matrices are to store information and to represent simultaneous linear equations. This appendix briefly reviews matrix conventions, matrix multiplication, and simple applications to simultaneous equations.

The values of a time series $x_t = [x_0, x_1, \ldots, x_{N-1}]$ need to be stored in an orderly way. This can be done either as a column vector

$$\begin{bmatrix} x_0 \\ x_1 \\ \cdots \\ x_{N-1} \end{bmatrix}$$

or as a row vector

$$\begin{bmatrix} x_0 & x_1 & \cdots & x_{N-1} \end{bmatrix}$$

The column vector is an $N \times 1$ matrix, and the row vector is a $1 \times N$ matrix. The two matrices are transposes of each other. If x is a row vector then x^T, its transpose, is a column vector. The relationship between a matrix and its transpose is that the first column of the matrix becomes the first row of its transpose, the second column becomes the second row of its transpose, and so on. When a matrix contains elements that are complex numbers, one can define the conjugate transpose, where the complex conjugate of the first column becomes the first row of the transpose, and so on. This is also called the Hermitian transpose or the adjoint. If we have several different time series, for example three time series $x1_t$, $x2_t$, $x3_t$ that are all the same length N, then it is convenient to store them as columns of a matrix.

Two vectors of length N have a dot (inner) product which is the natural extension of the dot product in two- or three-dimensional space. In ordinary three-dimensional space, when the dot product of two vectors is zero they are orthogonal and linearly independent. The concept extends to N-vectors as well: the dot or inner product is the sum

$$x \cdot y = x_0 y_0 + x_1 y_1 + \cdots + x_{N-1} y_{N-1}$$

The dot product is the essential operation that takes place during matrix multiplication. Each term at the position (row K, column L) in the resultant product matrix is the dot product between row K of the matrix on the left, and column L of the matrix on the right. Thus a dot product between two vectors can be expressed as matrix multiplication by taking x to be a $1 \times N$ row vector and y a $N \times 1$ column vector. The matrix product is then

$$\begin{bmatrix} x_0 & x_1 & \cdots & x_{N-1} \end{bmatrix} \begin{bmatrix} y_0 \\ y_1 \\ \cdots \\ y_{N-1} \end{bmatrix} = x_0 y_0 + x_1 y_1 + \cdots + x_{N-1} y_{N-1}$$

For matrix multiplication to work, the matrices must be compatible for multiplication, requiring that the number of columns of the matrix on the left equals the number of rows of the matrix on the right. The product is a new matrix with as many rows as the matrix on the left, and as many columns as the matrix on the right. That is, the product of matrix A of dimension $p \times q$ and matrix B of dimension $q \times r$ is a new matrix of dimension $p \times r$. Matrix multiplication is associative, but generally not commutative, and it is not even possible to change the order of matrix multiplication unless the matrices are square (they have same number of rows and columns). Even when this is the case, the product of commuted matrices is generally different.

Matrices and vectors are useful in representing and solving simultaneous linear equations. The basic principles can be illustrated by a two-dimensional example, solving two simultaneous linear equations. When the equations have two or three unknowns, there is a clear geometric interpretation. With additional unknowns, the geometry becomes obscure, but geometric terminology (such as orthogonality) extends to any higher dimension.

A linear equation in two variables describes a straight line in the plane of those two variables, and the simultaneous solution of two such equations is the intersection point of the two lines. Consider an (x, y) plane with two lines defined by the equations $y = x + 1$ and $y = -x + 1$. The lines have different slopes, but when $x = 0$ it is clear that both pass through $y = 1$, so the solution is $(x, y) = (0, 1)$. To formulate this in matrix notation, first rearrange the equations to put the two unknowns on the left:

$$-x + y = 1$$
$$x + y = 1$$

These equations can now be put into the standard form seen in many geophysical problems, where on the left we have the product of a coefficient matrix multiplying a column vector of unknown quantities, and on the right a column vector of numbers:

$$\begin{bmatrix} -1 & 1 \\ 1 & 1 \end{bmatrix} \begin{bmatrix} x \\ y \end{bmatrix} = \begin{bmatrix} 1 \\ 1 \end{bmatrix}$$

The product of a matrix and its inverse is the identity matrix, consisting of unit values on the diagonal and zeros off the diagonal. In this particular case, the inverse is proportional to the matrix itself, as we can show by factoring out $1/2$. This is not a common situation, and to find the inverse usually requires some effort. The solution to the equations is found by multiplying the left- and right-hand sides by the inverse of the matrix on the left. Here, we have for the inverse matrix

$$\begin{bmatrix} -1 & 1 \\ 1 & 1 \end{bmatrix}^{-1} = \frac{1}{2} \begin{bmatrix} -1 & 1 \\ 1 & 1 \end{bmatrix}$$

It can be seen that the inverse and the original matrix have the required property that

$$\frac{1}{2} \begin{bmatrix} -1 & 1 \\ 1 & 1 \end{bmatrix} \begin{bmatrix} -1 & 1 \\ 1 & 1 \end{bmatrix} = \begin{bmatrix} 1 & 0 \\ 0 & 1 \end{bmatrix}$$

Operating on the equation for the coefficient matrix with the inverse matrix provides the result that

$$\begin{bmatrix} x \\ y \end{bmatrix} = \begin{bmatrix} -1/2 & 1/2 \\ 1/2 & 1/2 \end{bmatrix} \begin{bmatrix} 1 \\ 1 \end{bmatrix} = \begin{bmatrix} 0 \\ 1 \end{bmatrix}$$

With only two equations to solve it is easy to obtain the inverse but, as the number of equations and thus the size of the matrix increases, considerable computational effort may be required. There are efficient algorithms for solving simultaneous linear equations that do not explicitly find the inverse of the corresponding matrix.

A matrix inverse is strictly defined only for square matrices, and even then an inverse may not exist. One might anticipate this situation if, in the example above, the two lines we had chosen were parallel. A simple example would be $y = x + 1$ and $y = x + 2$. Now there is no point of intersection, and no solution to the two equations. This becomes evident when the equations are put into matrix form:

$$\begin{bmatrix} -1 & 1 \\ -1 & 1 \end{bmatrix}\begin{bmatrix} x \\ y \end{bmatrix} = \begin{bmatrix} 1 \\ 2 \end{bmatrix}$$

The lack of an intersecting point is tied to the fact that the two columns of the coefficient matrix on the left are just a scalar multiple (-1) of each other. Thus they are linearly dependent. When two columns of a matrix are linearly dependent, then it has no inverse, and its determinant is zero.

A final interesting case occurs when the coefficient matrix is not square and so does not have an inverse in the usual sense. This arises when we are asked to find the intersection of three lines, rather than two. A third line will add a third row to the matrix. It is possible that an exact intersection point exists. That is, in a plane the three lines could intersect at a point. An interesting case is when the three lines almost intersect at a single point, but not exactly. An approximate result would be useful, if it corresponded to some sensible value near the three intersection points. The three-line intersection problem is over-determined, with more equations than unknowns, so there is generally not an exact solution. Chapter 7 examines least squares as a way to understand and solve such problems. For example, consider the three simultaneous equations

$$y = x + 1$$
$$y = -x + 2$$
$$y = 1$$

As simultaneous equations these can be represented in matrix form as

$$\begin{bmatrix} 0 & 1 \\ -1 & 1 \\ 1 & 1 \end{bmatrix}\begin{bmatrix} x \\ y \end{bmatrix} = \begin{bmatrix} 1 \\ 1 \\ 2 \end{bmatrix}$$

A solution can be found by multiplying the left- and right-hand sides by the transpose of the matrix on the left, yielding the *normal equations*, here two equations in two unknowns, which can be solved in the usual way. (The geometrical reasoning behind this approach to solving over-determined linear equations is found in Chapter 7, in the development of least squares.) The resulting column vector is

$$\begin{bmatrix} x \\ y \end{bmatrix} = \begin{bmatrix} 0.5 \\ 1.333 \end{bmatrix}$$

If you plot this (x, y) point along with the three straight lines that form the observation equations, you would confirm that there is no single intersection point of the three. There are three different points defining a triangular region. The solution obtained by reducing the three equations to two normal equations (after multiplying by the matrix transpose) falls within this triangular region and would be considered a reasonable approximation, even though an exact one does not exist.

Appendix B Fourier Transforms of Continuous Functions

Understanding the Fourier transform (FT) of a continuous function is essential to literacy in geophysics. Although geophysical data are almost entirely digital, the theory of the continuous function transform provides insight in designing filters and algorithms that will be eventually implemented in digital form. Beyond these data processing and analysis connections, the FT is the standard tool to analyze systems described by linear constant-coefficient differential equations, such as arise with seismometers and other geophysical devices. (The closely related Laplace transform is an alternative.) The FT also explains the sampling theorem and the phenomenon of aliasing, and shows how a continuous function can be recovered from its samples, effectively a proof of the sampling theorem.

The FT is closely related both to the DFT (Discrete Fourier Transform), used to analyze digital time series, and the Fourier series encountered in most college-level mathematics courses. We develop both connections below. The Fourier series provides a good introduction to the basic concepts of the FT, and we use that as the starting point. The DFT, really a discrete version of the Fourier series, is a relatively recent arrival in time series analysis literature, becoming prominent after the development of fast numerical algorithms (Fast Fourier Transforms) in the late 1960s. Given its relatively late arrival, many classical texts on Fourier transform theory either did not discuss the DFT or treated it as an afterthought. While aspects of FT theory involve considerable mathematical rigor, the DFT does not. In fact, the mathematics of the DFT only involves trigonometric functions and complex number addition, subtraction, and multiplication. Still, an understanding of the DFT and how to use it is enhanced by insights provided by the FT.

Many interesting physical processes are true continuous functions of time, so the FT of a continuous function is directly of interest in understanding the physics of such processes, even though we might eventually be dealing with discrete samples and need the DFT. Continuous functions in geophysics include the variation in sea level at a tide gauge, and the motion of the ground due to a passing seismic wave. Such a passing seismic wave is an example of a transient (that is, it goes to zero outside some time interval), with finite total energy, so the integral of the square of the function over all times is finite. Functions which are square integrable (have finite total energy) are mathematically suited to having a Fourier transform. On the other hand many geophysical variations persist for all time, so their energy (the integral of the squared variation about the mean) is infinite. Sea level variations are an example. Persistent time series have infinite energy, are not square integrable, and do not formally have a Fourier transform. However, this turns out not to be a problem in many applications. Functions that have infinite energy can still be analyzed in terms of their frequency content using Fourier transform concepts, appropriately modified to include the concept of the power spectrum. Thus, even though there are mathematical considerations that restrict Fourier transform theory to square integrable functions, the ideas are in fact generally useful in geophysics.

B.1 Fourier Series to Fourier Transform

We take as a starting point the complex Fourier series

$$x(t) = \sum_{m=-\infty}^{\infty} X_m \exp(2\pi i (m/N) t)$$

The expression for the coefficients is given by

$$X_m = \frac{1}{N} \int_{-N/2}^{N/2} x(t) \exp(-2\pi i (m/N) t) \, dt$$

Now consider the effect of the interval length N being increased while $x(t)$ does not change and, for example, could be assumed to be zero outside the original interval, of length N. This assures us that $x(t)$ is square integrable. As the interval grows larger, each frequency $f = m/N$ grows closer to its neighbors in the sum (the immediate neighbors are $(m + 1)/N$ or $(m - 1)/N$). This means, in the limit of N becoming infinite, that f takes on a continuum of values. The quantity $1/N$, the frequency difference between two neighbors in the series summation, can be called, in the sense of infinitesimal changes in calculus, a differential frequency df. We can then rewrite the expression for X_m in the limit as N goes to infinity as

$$df \int_{-\infty}^{\infty} x(t) \exp(-2\pi i f t) \, dt = X(f) \, df$$

The quantity $X(f) \, df$ has the same meaning as X_m as an amplitude of a sinusoid, so $X(f)$ by itself is the amplitude density or the amplitude per unit frequency and is defined as the Fourier transform of $x(t)$:

$$X(f) = \int_{-\infty}^{\infty} x(t) \exp(-2\pi i f t) \, dt$$

Turning now to the Fourier series sum itself, as the frequency f becomes continuous, integration can replace the sum on the right-hand side of $x(t)$ to give

$$x(t) = \int_{f=-\infty}^{\infty} X(f) \exp(2\pi i f t) \, df$$

These two equations give the formal definitions of the Fourier transform (finding $X(f)$ from $x(t)$) and of its inverse (finding $x(t)$ from $X(f)$). The general symmetry of the forward and inverse transforms is remarkable: they differ only in the sign of the complex exponential. The Fourier transform $X(f)$ is complex even if $x(t)$ is real, but the theory is applicable to complex functions in general, so both $x(t)$ and $X(f)$ may be complex-valued.

B.2 Fourier Transform Notation and the Role of Symmetry

As shorthand notation, $F[]$ denotes the operation of Fourier transformation, that is, $F[x(t)] = X(f)$ and $F^{-1}[X(f)] = x(t)$. The convention of using capital and lower case letters to indicate transform pairs is common, but there are others. Some authors use a circumflex to denote the transform, while a few use exactly the same letter and let either the context, or the variable, indicate which is the transform. As a transformation, the FT is linear, so the transform of the sum of two functions is the sum of their transforms. While the definition that we have used (regarding the sign of the exponentials, the use of frequency f as the variable) is common, there is also some variety within the literature. In certain cases, the sign of the complex exponential is interchanged between the forward and inverse transform definitions, with the inverse transform receiving the negative sign in the exponential. This is common in elastic-wave propagation theory, where Fourier transforms with respect to spatial coordinates are given one sign for the exponential and the time transforms then have the opposite sign. Another common difference is to use angular frequency $\omega = 2\pi f$ in place of f, which requires introducing factors of 2π here and there due to this change of variable.

The existence of a Fourier transform pair, that is, $x(t)$ and $X(f)$, implies two alternative but equivalent descriptions of a function $x(t)$. One, the function itself, is the time domain description and the other, $X(f)$, is the frequency domain description. The forward transform (time to frequency) determines the amplitude density of a complex sinusoid of frequency f in the time function. The amplitude at frequency f is the product $X(f)df$. The inverse transform (frequency to time) is an integration of complex sinusoids of various frequencies to form the time function. For the usual case when $x(t)$ is purely real, $X(f)$ will generally be complex, but will have Hermitian symmetry, meaning that $X(-f)$ is the complex conjugate of $X(f)$. We can form a real-valued function $|X(f)|$ which is an even function of f (for real $x(t)$) known as the amplitude spectrum.

To investigate the role of Hermitian and other symmetry in Fourier transform theory, we observe that any function $x(t)$ defined on the infinite interval can be separated into an even function $e(t)$ symmetric about $t = 0$, and an odd function $o(t)$ in the following way:

$$e(t) = [x(t) + x(-t)]/2$$
$$o(t) = [x(t) - x(-t)]/2$$
$$x(t) = e(t) + o(t)$$

Clearly $e(-t) = e(t)$ and $o(-t) = -o(t)$.

The importance of an even–odd separation is that a complex exponential includes both a cosine (even) and sine (odd) function. Because integrating the product of an even and odd function on a symmetric interval produces a zero result, we conclude that

$$X(f) = 2 \int_0^\infty e(t) \cos(2\pi ft)\, dt + (2i) \int_0^\infty o(t) \sin(2\pi ft)\, dt$$

Thus, the real part of the Fourier transform is twice the cosine transform of the even part of the function, and the imaginary part is twice the sine transform of the odd part. If a function is even then its Fourier transform is pure real, and if odd, then it is pure imaginary.

From the Fourier transform definition we can conclude a few other things related to symmetry. One is that the Fourier transform of $x(t)$ reversed in time, denoted as $x(-t)$, is the complex conjugate of $X(f)$, because the effect of time reversal is just to change the sign of the odd part. For any real-valued function, the imaginary part of its Fourier transform must be odd. That is, if we replace f with $-f$ then the real part is unchanged (because cosine is an even function) but the imaginary part reverses sign. A function that has such a property (even real part, odd imaginary part) is Hermitian symmetric.

Hermitian symmetry is also a feature of the DFT, but it is obscured by the conventional way in which the DFT is defined on the time interval $[0, N - 1]$. However, the last half of a time series on $[0, \ldots, N - 1]$ may also be interpreted as negative times. That is, the time $t = N/2$ can also be interpreted as $t = -N/2$, and so on up to $t = N - 1$, which can be interpreted as $t = -1$. This is so because the DFT "thinks" that every time series defined on the interval $[0, N - 1]$ is periodic with period N. Therefore, we can choose any interval to compute the DFT, and this was done earlier when we changed the definition to be symmetric about time zero. To clarify this, consider some examples. First, a four-point time series whose times are usually give as $t = [0, 1, 2, 3]$ can also be interpreted as $t = [0, 1, \pm 2, -1]$. Thus, for the DFT, an example of a time-symmetric (even) time series is the series $[1, 1, 0, 1]$. The DFT of this series is purely real, just as the Fourier transform of a symmetric time function would be. Similarly, an odd time series (as seen by the DFT) would be $[0, -1, 0, 1]$, for which the DFT is pure imaginary, just as the Fourier transform of an odd function is pure imaginary. In the frequency domain the dual interpretation of frequency values is similar to that in the time domain. That is, the frequency at $m = N - 1$ which is $f = (N - 1)/(N\Delta t)$ is the same as $f = -1/(N\Delta t)$. Thus negative frequencies appear in the DFT array in reverse order, after the positives, and the DFT values at the positive and negative Nyquist frequencies are the same.

B.3 Convolution and Correlation

Convolution and correlation are operations between two time functions that define a new function of time. The two operations are closely related, and we begin with a discussion of convolution. We assume each has a Fourier transform, and present a convolution theorem that relates the operations of convolution in the time and frequency domains.

The convolution of continuous functions is defined in terms of multiplication and integration, in a way analogous to the operations of multiplication and summation in discrete convolution. If $g(t)$ and $h(t)$ are two continuous functions, then their convolution is a new function of time t, defined as

$$g(t) * h(t) = \int_{-\infty}^{\infty} g(u)h(t - u)du$$

Here u, as the variable of integration, disappears after the integral is carried out, and time reversal of one function is evident from the appearance of $h(t - u)$ in the integrand. This is the continuous function version of discrete convolution, and by simple changes of variable one can show that this operation between two functions is commutative, associative, and distributes over addition. That is, for continuous functions $g(t), h(t), k(t)$ (dropping the argument t), $h * g = g * h$ (commutative); $h * (g + k) = h * g + h * k$ (distributive); and $h * g * k = (h * g) * k$ (associative).

The convolution theorem relates the operation of convolution in the time domain to multiplication in the frequency domain. The theorem states that the Fourier transform of the product of two functions is the convolution of their Fourier transforms and the Fourier transform of the convolution of two functions is the product of their transforms. The convolution theorem applies equally to the time or frequency domains, owing to the symmetry of the transforms. The proof, in the case of continuous functions, is obtained by directly substituting the convolution integral into the Fourier transform integral:

$$F[g(t) * h(t)] = G(f)H(f)$$

and, because of the symmetry of the forward and backward transforms, the following is also true:

$$F[g(t)h(t)] = G(f) * H(f)$$

These results parallel the properties of discrete convolution and its relationship with the Z polynomial formed from it (its Z transform). That is, the Z transform of the discrete convolution of two time series is the product of the Z transform of each series. This connection with polynomial multiplication guarantees that discrete convolution has the same associative, commutative, and distributive properties as ordinary multiplication. In the case of the DFT, the convolution theorem for discrete time series must be modified in one of two ways. Either the definition of convolution is modified to cyclic convolution, or the lengths of all time series are extended, by padding zeros, to be the same.

Cross correlation is similar to convolution, except that there is no reversal in time. As a result, cross correlation is not a commutative operation. The operation is defined using a pentagram symbol $\star$, as follows:

$$g(t) \star h(t) = \int_{-\infty}^{\infty} g(u)h(t+u)du$$

Cross correlation is therefore convolution between one function, and a second function which has first been time reversed.

There are important applications of cross correlation in data processing. An example of a common application is to "find" a copy of $h(t)$ that is contained somewhere in the time function $g(t)$ at an unknown time. This is done by computing the cross correlation function and noting where it has a locally maximum value. The cross correlation operation involves sliding $h(t)$ past $g(t)$ ($h(t)$ is not time reversed), multiplying, and then integrating. The variable t in the cross correlation of $g(t)$ and $h(t)$ has the meaning of the lag or shift between the two functions. If there are wiggles in $g(t)$ similar to $h(t)$ then when the lag or shift causes the two similar features to overlap, their product will be large, and the result will be similar to the square of $h(t)$. When integrated, this all-positive squared function will produce a large value of the cross correlation. Cross correlation is used to improve signal to noise performance in radar, sonar, and Vibroseis. The cross correlation theorem is

$$F[g(t) \star h(t)] = G(f)H^*(f)$$

A special case is that of autocorrelation, in which a function is cross-correlated with itself. The result is a new function derived from the original, called the autocorrelation function. The autocorrelation function is an even function of lag, or in the complex case it is Hermitian-symmetric. Thus the autocorrelation function has a real-valued Fourier transform equal to the squared modulus of the transform, that is,

$$F[g(t) \star g(t)] = \|G(f)\|^2$$

The squared modulus of the Fourier transform is called the energy spectrum, and describes how energy is distributed over frequency. Obviously, the energy spectrum does not uniquely define the function $g(t)$, since many functions may have the same energy spectrum or, equivalently, many different functions may possess the same autocorrelation function.

The Rayleigh theorem is a statement about how the integral of the energy spectrum over all frequencies is related to the integrated variance over all time. The Rayleigh theorem states that

$$\int_{-\infty}^{\infty} \|g(t)\|^2 dt = \int_{-\infty}^{\infty} \|G(f)\|^2 df$$

The analogous result is the Parseval theorem for the DFT:

$$\sum_{t=0}^{N-1} \|x_t\|^2 = (1/N) \sum_{m=0}^{N-1} \|X_m\|^2$$

The Rayleigh theorem is the starting point for discussions of the power spectrum, and the Parseval theorem for the DFT is the starting point for developing practical algorithms to estimate the power spectrum from time series.

B.4 Similarity, Shift, and Derivative Theorems

From the symmetry of the Fourier transform, any theorem stated for time domain operations and their consequences in the frequency domain may be restated for the corresponding frequency domain operation and consequences in the time domain.

The similarity theorem is

$$F[g(at)] = \frac{1}{|a|}G(f/a)$$

where the scalar a is an unspecified constant. This shows how stretching in one domain produces compression (shrinking) in the other. For example, if the parameter a is larger than 1 in magnitude, the function $g(at)$ is compressed on the time axis, relative to $g(t)$, but in the frequency domain $G(f/a)$ is stretched to include a wider range of frequencies, that is, more high frequencies. A compressed time function $g(at)$ requires more high frequencies to make it up, hence we have a compression–stretching relationship between the two domains.

The shift theorem is

$$F[g(t - a)] = \exp(-2\pi i a f)\, G(f)$$

A shift in time changes only the phase, not the amplitudes of the sinusoids making up the time function $g(t)$. The magnitude of $|\exp(-2\pi i a f)|$ is unity for all real f and a, and the phase change is a linear function of frequency for a simple shift in time given by the parameter a. The shift theorem is implemented in computation by finding the DFT of a time function, multiplying by the complex exponential shift factor, and then computing the IDFT.

The derivative theorem shows how to find the Fourier transform of the time derivative of a function. Letting $g'(t)$ denote the first derivative of g with respect to t, we have

$$F[g'(t)] = i2\pi f G(f)$$

Repeated application of the derivative theorem leads to the result that the second derivative has the transform

$$F[g''(t)] = -(2\pi f)^2 G(f)$$

and the Fourier transform of each higher derivative is found by multiplying by an additional factor $2\pi i f$. This result is of fundamental importance in analyzing seismic wave fields, which are solutions to linear partial differential equations (wave equations), as well as in understanding physical devices governed by differential equations.

The derivative theorem is useful in understanding the frequency domain behavior (response) of physical systems governed by differential equations that are linear with constant coefficients. A simple harmonic oscillator (for example, a seismometer) is a common application in geophysics, where the governing equation is of the form

$$a\frac{d^2 y(t)}{dt^2} + b\frac{dy(t)}{dt} + c y(t) = x(t)$$

The constants (a, b, c) can be related to the parameters of the physical system, such as spring constants, mass, damping, and other parameters. For a seismometer, for example, $x(t)$ might represent the ground acceleration, while $y(t)$ might represent the displacement of the seismometer mass–spring system. The system response is the ratio at each frequency of the output divided by the input. Taking the Fourier transform of both sides and rearranging, one obtains the system transfer function, denoted by $L(f)$. The concept parallels the idea of the transfer function for linear digital filters. The result is

$$\frac{Y(f)}{X(f)} = L(f) = \frac{1}{a(2\pi i f)^2 + b(2\pi i f) + c}$$

The function $L(f)$ is called the transfer function and describes how each frequency of the input $x(t)$ is amplified (or attenuated) and phase shifted, via the complex number $L(f)$ at frequency f. The output can be found from the input via the inverse transform:

$$y(t) = \int_{-\infty}^{\infty} L(f)X(f)\exp(2\pi i f t)\, df$$

which is a summation of the input at each frequency $X(f)$, scaled by the transfer function, to obtain the output amplitude $Y(f)$.

The derivative of a convolution theorem is obtained by application of the convolution theorem. If $g(t)$ and $h(t)$ are two functions and $g(t) * h(t)$ is their convolution, another function of time, then the time derivative of this convolution is the convolution of the time derivative of either function with the other. That is,

$$[g(t) * h(t)]' = g'(t) * h(t) = g(t) * h'(t)$$

The proof of this is found by taking the Fourier transform of the derivative, and noting that in the frequency domain the multiplying factor $2\pi i f$ can be assigned to either $G(f)$ or $H(f)$.

B.5 Boxcar and Sinc Functions

There are a number of special functions in Fourier transform theory for which the Fourier transform is easily calculated. Two such functions, the boxcar and sinc functions, form a Fourier transform pair. The boxcar function $\Pi(t)$ is defined to be zero except on the interval $[-1/2, 1/2]$, where it has the value 1. One may verify by direct integration that

$$F[\Pi(t)] = \frac{\sin(\pi f)}{\pi f} = \text{sinc}(f)$$

This defines the sinc function. One can quickly deduce that

$$F[\text{sinc}(t)] = \Pi(f)$$

That is, the boxcar and sinc functions are Fourier transform pairs. It is useful to consider an interpretation of the $\text{sinc}(f)$ function as an expression of the frequency content of the boxcar function. The discontinuities in the boxcar function require that it be constructed from a sum of sinusoids over an infinite frequency range. Thus $\text{sinc}(f)$ never goes exactly to zero. On the other hand, $\text{sinc}(t)$ is band-limited, since its transform $\Pi(f)$ goes completely to zero at frequency one-half and higher frequencies. This behavior can be summarized by saying that time-limited functions are not band-limited, while band-limited functions are not time-limited.

With one Fourier transform pair (boxcar–sinc), we can generate others using the theorems stated earlier. For example, we define the triangle function $\Lambda(t)$ as

$$\Lambda(t) = \Pi(t) * \Pi(t)$$

The function $\Lambda(t)$ is zero outside the interval $[-1, 1]$ and rises linearly from $t = -1$ to $t = 0$, where it achieves the peak value of 1 and then falls linearly to 0 at $t = 1$. The convolution gives us immediately that

$$F[\Lambda(t)] = \text{sinc}^2(f)$$

and provides the other Fourier transform pair,

$$F[\text{sinc}^2(t)] = \Lambda(f)$$

There are many other continuous functions for which Fourier transform pairs can be deduced, but the boxcar and sinc are sufficient for us to proceed with a discussion of the most important of the special functions, the Dirac delta function.

B.6 The Delta Function

The delta sequence $\delta_t = [1, 0, 0, \ldots]$ appears in our discussion of discrete time series. It is recognized as a time series that is the identity operator for discrete convolution, and whose DFT is a constant at all frequencies. These properties are completely analogous to those of the Dirac delta function $\delta(t)$, though the delta sequence requires special mathematical consideration. The delta function is defined to be zero for all t, except at $t = 0$, where it is infinite. The infinite value means that it is not square integrable. Still, it has a Fourier transform, which is well behaved. It is defined to have unit area, contained within the infinite spike at $t = 0$:

$$\int_{-\infty}^{\infty} \delta(t) dt = 1$$

The Fourier transform of the delta function is

$$F[\delta(t)] = 1$$

That is, the delta function contains all frequencies in equal amounts. The other transform pair that follows at once is

$$F[1] = \delta(f)$$

This means that a constant function in the time domain has only one frequency constituent, namely $f = 0$, since $\cos(2\pi f) = 1$ when $f = 0$.

To demonstrate that the Fourier transform of $\delta(t)$ is 1 is surprisingly easy. The approach is to recognize that $\delta(t)$ is a limit of a sequence of functions. Consider the following sequence of boxcar functions as the positive constant a is allowed to approach zero:

$$(1/a)\Pi(t/a)$$

The boxcar functions in the sequence become narrower and taller, but preserve unit area. As the parameter a goes to zero, the functions are not square integrable, because the integral of the square is $1/a$, which becomes infinite.

The limit of the sequence of functions has all the properties we ascribe to the Dirac delta, $\delta(t)$. It is infinite at one point, has unit area, and is zero elsewhere. To find the Fourier transform note that the similarity theorem gives each member of the sequence of functions a transform $\text{sinc}(af)$. Thus, the limit when $a = 0$ produces a transform $\text{sinc}(0) = 1$. This verifies the claim that the delta function transform is well behaved, though not square integrable.

Any suitable function can be used to form the sequence of functions. Examples include those developed above (sinc, triangle, and boxcar) as well as many others. The Dirac delta is considered to be a generalized function because of its definition as a limit of a sequence of functions.

Generally, the properties of the delta function can be derived using a sequence of functions (for example, boxcar functions as above), and taking the limit. An example is the sifting property:

$$x(t) = \int_{-\infty}^{\infty} x(u)\delta(t-u)du$$

The function $\delta(u-t)$ is peaked at the value $u = t$, and so selects out the value of $x(u)$ at $u = t$.

B.7 The Sampling Function and the Sampling Theorem

The sampling theorem sets the conditions under which a continuous function may be recovered from its samples. The requirement is that when a function $x(t)$ is sampled at a uniform rate there must be at least two samples of the highest frequency present in $x(t)$. This statement is usually given without proof. The consequence of undersampling is that then $x(t)$ cannot be recovered from its samples. This means that not only do frequencies in $x(t)$ above the Nyquist get lost but aliasing, a confusion of frequencies, occurs. Aliasing arises because samples of sinusoids above the Nyquist are identical to samples of frequencies within the Nyquist band.

Thus the sampling theorem states that a continuous-time function can be completely recovered from its samples taken at intervals of Δt provided that no frequency larger than $f_{Nyquist} = 1/(2\Delta t)$ is present in the function.

To show how to reconstruct the continuous function, we need to define the sampling function, also called the Dirac comb:

$$III(t) = \sum_{n=-\infty}^{+\infty} \delta(t-n)$$

where the sample time n takes on integer values $[-\infty, \ldots, -1, 0, 1, \ldots, +\infty]$. The sampling function is a set of delta functions at unit spacing on the time axis. Using sequences of integer-spaced functions (such as boxcars) in the limit in which they become integer-spaced delta functions, one can show that the Fourier transform of the sampling function is given by $F[III(t)] = III(f)$. Thus, the sampling functions in time and frequency form a Fourier transform pair.

Uniform sampling of a function $x(t)$ produces an infinite set of samples, assumed to be instantaneous values of the continuous function:

$$[\ldots, x(-1), x(0), x(1), \ldots] = [\ldots, x_{-1}, x_0, x_1, \ldots]$$

This infinite set of samples determines the function $III(t)x(t)$, which is the product of the continuous function $x(t)$ and the sampling function. The Fourier transform of this product of two functions is given by the convolution theorem:

$$F[III(t)x(t)] = III(f) * X(f)$$

Convolution in the frequency domain with $III(f)$ has the effect of replicating $X(f)$ periodically on the infinite frequency axis, providing that $X(f)$ goes to zero outside the range $f = [-1/2, 1/2]$. If this condition is violated, that is, if $X(f)$ is non-zero outside the frequency band $f = [-1/2, 1/2]$, then the effect of convolution with $III(f)$ is to smear out $X(f)$ in such a way that there is a periodically replicated function, but it is not recognizable as repeated copies of $X(f)$. These repeated copies appear at other frequencies, which are the alias frequencies of those in the range $f = [-1/2, 1/2]$. It is clear that if $x(t)$ is not band-limited then these alias frequencies will appear within the band of the principal alias $f = [-1/2, 1/2]$, and become confused with values in this range. Of course, if the principal alias band is not $f = [-1/2, 1/2]$, but any other range of frequencies of unit length, then there is still no problem with aliasing. $X(f)$ would still be band-limited, but limited instead to some band outside the usual principal Nyquist range.

To recover $x(t)$ from its samples, we select the principal alias from $III(f) * X(f)$, which is just the part of this function between $[-1/2, 1/2]$. This is done by multiplying the periodically replicated function by a boxcar function:

$$X(f) = \Pi(f)[III(f) * X(f)]$$

Again using the convolution theorem to go back to the time domain, we conclude that

$$x(t) = \text{sinc}(t) * [III(t)x(t)]$$

This last expression can be evaluated directly since convolution with a delta function sifts out values of the sinc function. The final result is Whittaker's interpolation formula:

$$x(t) = \sum_{n=-\infty}^{+\infty} x(n) \,\text{sinc}(t - n)$$

This is an explicit statement showing how to find $x(t)$ for any t as a linear combination of samples taken at integer times. The sinc function weights diminish in amplitude as n increases, so samples near the time of interest get the largest weight but all samples contribute, and their associated coefficients oscillate in sign. The Whittaker formula is not practical because it requires an infinite set of terms, but it does provide a proof of the sampling theorem.

This analysis allows us to understand what frequencies will appear as aliases. The principal alias is (normally) in the band $[-1/2, 1/2]$, but all frequencies separated by unit frequency $\Delta f = 1$ (the sampling frequency) on the frequency axis are aliases of the principal alias. That is, for f in the interval $[-1/2, 1/2]$, its alias frequencies are $[\ldots, f - 2, 1, f, f + 1, f + 2, \ldots]$. Providing $X(f)$ is zero outside the Nyquist band, aliasing is not a problem. As an example, suppose that we are digitizing a voltage with a sampling frequency of 9 Hz, so the Nyquist frequency is 4.5 Hz. Suppose that 60 Hz power line voltages are not removed by the anti-alias filter. What frequency will 60 Hz appear to be after it is sampled? Considering both positive and negative 60 Hz, the alias of +60 Hz is at

$60 - 7 \times 9$ Hz $= -3$ Hz and the alias of -60 Hz is at the frequency $-60 + 7 \times 9$ Hz $= +3$ Hz. That is, sampling a 60 Hz sinusoid at nine samples a second will produce samples that look just like those of a 3 Hz sinusoid. In this way, 3 Hz is the alias of 60 Hz.

B.8 Power Spectrum of a Continuous Function

The Rayleigh theorem describes how total energy in the time domain is distributed over frequency:

$$\int_{-\infty}^{\infty} \|x(t)\|^2 dt = \int_{-\infty}^{\infty} \|X(f)\|^2 df$$

On the left, the units are variance (square of x) multiplied by time, the total energy of the time function. On the right $\|X(f)\|^2$, with units of energy per frequency, is the energy spectrum. From Fourier transform theory (Section B.3) the energy spectrum is also the Fourier transform of the autocorrelation $\rho_x(t)$ of $x(t)$:

$$F[x(t) \star x(t)] = F\left[\int_{-\infty}^{\infty} x(u)x^*(u+t)du\right] = F[\rho_x(t)] = \|X(f)\|^2$$

Fourier transform theory (and the Rayleigh theorem) apply only to square integrable functions with finite total energy, but many geophysical quantities (sea level variation is an example) have infinite energy (the integral of the square of the function over time tends to infinity). Therefore the sea level time function does not have a Fourier transform, even though it contains interesting variations with distinct time scales (wind waves at periods of seconds; tides at periods near 12 and 24 hours; and Milankovitch periods of 20-, 40-, and 100-thousand years). To adapt the Rayleigh theorem to deal with such series, we introduce the idea of power, the rate of energy delivered per unit time. The frequency domain counterpart is the power spectrum. Power is what we have previously called variance, the average squared variation. For geophysically interesting quantities, power (variance) is finite. To adapt the Rayleigh theorem (Section B.3) to this new quantity, replace the left-hand side with averages over intervals of increasing duration. On the right-hand side, the energy spectrum is replaced by its average over time, with units of variance per frequency. It can be called either the power spectrum or the power spectrum density.

The mean squared value of $x(t)$ is the power in $x(t)$. It is the average squared value of $x(t)$ over a time interval T in the limit as T becomes infinite:

$$\lim_{T \to \infty} \frac{1}{T} \int_{-T/2}^{T/2} \|x(t)\|^2 dt$$

Its Fourier transform becomes, in the limit of infinite T,

$$\lim_{T \to \infty} \frac{\|X(f)\|^2}{T} = P_x(f)$$

where $P_x(f)$ is the power spectral density, with units of variance per frequency.

The connection between the power spectrum and the autocorrelation function is also modified. The average autocorrelation in the limit of increasing intervals T is

$$r_x(t) = \lim_{T \to \infty} \frac{1}{T} \int_{-T/2}^{T/2} x(u)x^*(u+t)du$$

The revised autocorrelation at zero lag is the mean square value, equal to the variance. The new definition of the autocorrelation is related to the power spectrum by

$$F[r_x(t)] = P_x(f)$$

These modifications of the Rayleigh theorem were presented in the 1930s and 1940s by Wiener and Khintchine.

B.9 The Sign and Heaviside Functions, and the Hilbert Transform

The Heaviside function $H(t)$ is the indefinite integral of the delta function,

$$H(t) = \int_{-\infty}^{t} \delta(u)du$$

where u is a variable of integration. The function $H(t)$ is zero for negative t, unity for positive t, and discontinuous at $t = 0$. It is also called the unit step function, and provides an alternative to the delta function as a way to test the response of a linear system. Related to the Heaviside function is the sign function, denoted sgn(t), and defined to be $+1$ for positive t, and -1 for negative t. That is, sgn(t) just gives the value 1 multiplied by the sign of the variable t.

The Fourier transform of sgn(t) can be found using the transform integral and contour integration in the complex plane, topics not of our concern here. The important point is the result:

$$F[\text{sgn}(t)] = -i\frac{1}{\pi f}$$

Note that, since sgn(t) is odd, its transform must be pure imaginary, with the negative sign arising from the sign definition of the forward transform $\exp(-2\pi i f t)$. Thus the reciprocal relationship must be that

$$F^{-1}[\text{sgn}(f)] = +i\frac{1}{\pi t}$$

because the sign of the exponent is positive for the inverse transform. We can immediately use this result to write down the Fourier transform of the Heaviside step function of time. Expressing the step function as

$$H(t) = \frac{1}{2} + \frac{\text{sgn}(t)}{2}$$

and noting that the transform of the sum of two functions is the sum of their transforms, we obtain

$$F[H(t)] = \frac{1}{2}\delta(f) - \frac{1}{2}\frac{i}{\pi f}$$

From the symmetry of the tranform we can also write down the reciprocal relationship for frequency and time. Since $2H(f) = 1 + \text{sgn}(f)$, and noting (as above) the sign change on the inverse transform of sgn(f), we obtain

$$F^{-1}[2H(f)] = \delta(t) + \frac{i}{\pi t} = \delta(t) - i\left(\frac{-1}{\pi t}\right)$$

With these preliminaries, we introduce a new concept, the Hilbert transform of a function. For a function $x(t)$ the Hilbert transform is defined to be the convolution

$$Hi[x(t)] = \frac{-1}{\pi t} * x(t) = x_{hi}(t)$$

The Hilbert transform of a function $x(t)$ is its 90-degree-phase-shifted version, also called the quadrature of $x(t)$.

The analytic signal related to $x(t)$ is defined as

$$x(t) - ix_{hi}(t) = \left[\delta(t) - i\left(\frac{-1}{\pi t}\right)\right] * x(t)$$

The imaginary part of the analytic signal is the negative of the Hilbert transform. In the frequency domain, the convolution operation becomes multiplication by the function $2H(f)$. Thus the analytic signal contains all the original information in a real-valued $x(t)$, since it includes all the positive frequencies, and the negatives contain no separate information (they are just complex conjugates of the positives). The Fourier transform of the analytic signal is zero for negative frequencies, and for positive frequencies it is

$$F[x(t) - ix_{hi}(t)] = 2X(f)$$

A simple example will illustrate the connection between a function and its Hilbert transform. The function

$$\cos(2\pi ft) = \frac{\exp(2\pi ift) + \exp(-2\pi ift)}{2}$$

has in the frequency domain a contribution from both positive and negative frequencies. If the negative-frequency portion is removed by multiplication by $2H(f)$ then the result in the time domain is the analytic signal derived from the cosine function, which is just twice the positive-frequency part of the cosine function. That is, we conclude that

$$\left[\delta(t) - i\left(\frac{-1}{\pi t}\right)\right] * \cos(2\pi ft) = \exp(2\pi ift)$$

so the Hilbert transform of the cosine is

$$Hi(\cos(2\pi ft)) = -\sin(2\pi ft)$$

The fact that the analytic signal (hence the Hilbert transform) can be obtained by frequency domain multiplication by the Heaviside function suggests an easy algorithm for computing the analytic signal for discrete time series. We can use the DFT to take us to the frequency domain, double the positive frequencies, set to zero the negative-frequency values, leave zero and the Nyquist frequencies unchanged, then transform back to the time domain.

The Hilbert transform has important applications in data processing. One is to find the analytic signal and the related instantaneous amplitude and instantaneous frequency. Consider a pure funtion $\cos(2\pi ft)$, with a single frequency, whose 90-degree-phase-shifted version is $\sin(2\pi ft)$. A function and its Hilbert transform multiplied by the imaginary unit i can be combined to form the analytic signal, in this case $\cos(2\pi ft) + i\sin(2\pi ft) = \exp(i2\pi ft)$. From the analytic signal we can define an instantaneous amplitude which is the modulus of the analytic signal, and equal to unity in this case and an instantaneous phase $2\pi ft$. The time derivative of the instantaneous phase (divided by 2π)

is called the instantaneous frequency, in this case simply f. While these results are trivial for a simple sinusoid, they can be extended to an arbitrary time series, allowing us to determine these two instantaneous quantities. The interpretation of the instantaneous amplitude is that it is the envelope function for an oscillatory signal, and the instantaneous frequency describes how the dominant frequency changes over time. A computational algorithm to find the analytic signal using the DFT is given in Section 4.9.

Appendix C Random Variable Concepts and Applications

This appendix contains an abbreviated discussion of the random variable concepts that underly least squares and other methods of data processing and analysis. The concept of a random variable (abbreviated rv) shows up frequently in geophysical data processing and analysis. For example, noise or measurement error is a presumed contaminant in every measurement and is dealt with using random variable concepts. Any problem in which there is uncertainty is also suitable for random variables. In geophysical problems, uncertainty in estimates may be due to noise in the data, to insufficient data, or to errors in the theory or model relating the data and the quantities to be estimated. Probability and random variable concepts provide a useful framework to quantify uncertainty in estimates, regardless of the cause.

C.1 Probability Density Functions

Suppose we have a geophysical time series corrupted by a noise time series n_t that consists of some type of random numbers. For the moment, assume that the noise from one sample to the next is of the same general size, but independent, in the sense that there is no connection between the noise in one sample and that in the next or any other sample. This is called stationary and white noise. Stationary refers to the assumption that the noise does not change its properties over time, so its statistics, such as mean and variance, remain constant. The term white is associated with independence from sample to sample. This makes its autocorrelation tend to be zero except at zero lag, so its power spectrum, the Fourier transform in the continuous case (or the DFT in the discrete case) is flat, with equal contributions from all frequencies.

A description of the noise associated with the time series is given by a continuous function $p_n(u)$, the probability density function, which is a complete description of the rv that appears in each value of the time series n_t. The subscript n identifies $p_n(u)$ with the rv in n_t. The variable u is allowed to take on any possible value in n_t. The pdf gives the probability density that a value of n_t will be found within a certain range. To find the probability that n_t will take on a value somewhere between u_1 and u_2, one computes the area under the pdf between those limits:

$$\int_{u_1}^{u_2} p_n(u)\,du$$

If u_1 is the smallest value that n_t can have (it could be $-\infty$) and u_2 is the largest value possible (it could be $+\infty$) then the integral, the area under the pdf, has a value of unity. The fact that the area under a pdf is unity means that it is certain that the rv will take on some value within its allowed range. In the remaining discussion, we assume a range $(-\infty, +\infty)$.

It is common to summarize a pdf using a few important parameters derived from the continuous function. The first of these is the expected value, also called the mean value μ_n. Conceptually, the expected value is the average of many realizations of the rv. Thus, the mean or expected value of u is found by integration:

$$E[n] = \mu_n = \int_{-\infty}^{+\infty} u p_n(u) du$$

The expected value of any function $g(n)$ of the random variable n, also called the expectation of $g(n)$, is

$$E[g(n)] = \int_{-\infty}^{+\infty} g(u) p_n(u) du$$

where the pdf is used as a weighting function in the integration. A very important point is that, even if we don't know the pdf, the expectation is a linear operator (because integration is a linear operation). This linearity allows us to come to important conclusions without knowledge of the specific pdf. We use the expectation operator $E[\,]$ in this way elsewhere.

Another parameter of frequent interest is the second moment about the mean, called the variance. This is the expected value of the square of the difference between the random variable and its mean:

$$E[(n - \mu_n)^2] = \sigma_n^2 = \int_{-\infty}^{+\infty} [u - \mu_n]^2 p_n(u) du$$

The standard deviation σ_n is the square root of the variance. There are higher-order moments of random variables, so the kth moment of a random variable about its mean is

$$m_n^k = \int_{-\infty}^{+\infty} [u - \mu_n]^k p_n(u) du$$

The first moment about the mean is zero. The third moment about the mean is called the skewness, and the fourth moment is the kurtosis. Higher-order moments are rarely discussed, though they are of theoretical interest because the full (infinite) set of moments of a pdf completely determines it.

In addition to the mean and variance, two other pdf parameters are useful. One is the median value. The area under the pdf to the left and right of the median is exactly 0.5, so there is an equal chance (50 percent probability) that an rv will have a value larger or smaller than the median. The other common parameter(s) is (are) the mode(s) of the pdf, defined to be the values of u where there is a local peak. Pdfs with a single mode are called unimodal, those with two modes are bimodal, and those with more are multi-modal. Modes are regions of high probability density. For some pdfs the median, mean, and mode are all the same. In other cases, there may be no mode, or the mode, mean, and median may all differ.

The parameters of a pdf, including mean, variance, standard deviation, and median are distinct from the statistics computed from a particular data set which have the same names. Thus the statistics from a particular data set (which can be taken as estimates of the pdf parameters) are called the sample mean, sample variance, and so on, if there is a possibility of confusion.

C.2 Three Important PDFs

So far we have said nothing about the pdf itself, so it could be any function at all, providing the area under it is unity, and it is non-negative (since negative probabilities do not make sense). There

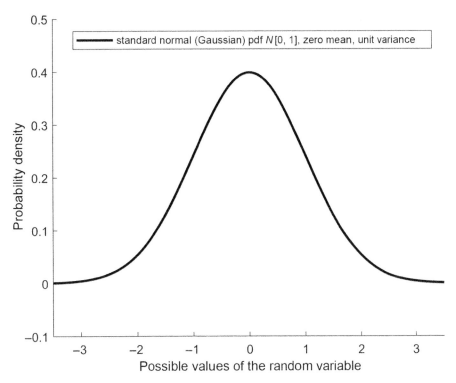

The Gaussian (normal) pdf for the standard case of zero mean and unit variance. About two-thirds of the area between the curve and the zero probability density line is within the range $[-1, 1]$ or plus and minus one standard deviation of the mean.

are only a few types of random variables (pdfs) in common geophysical use. We describe briefly properties of three, the normal (Gaussian), uniform, and chi-squared distributions.

C.2.1 The Gaussian or Normal PDF

The most important pdf is the normal distribution, also called the Gaussian, after the nineteenth century scientist Karl F. Gauss (see Figure C.1). It has remarkable properties, is used in experimental and other sciences, and is commonly assumed when one does not actually know much about a random variable. Its pdf is the function

$$p_n(u) = \frac{1}{\sqrt{2\pi\sigma_n^2}} \exp\left(-\frac{1}{2}[(u - \mu_n)/\sigma_n]^2\right)$$

which describes a bell-shaped curve that is symmetric about μ_n. The width of the bell is determined by σ_n, with a larger value producing a fatter bell curve. The area under the curve is unity, regardless of the values of parameters μ_n and σ_n. The Gaussian pdf is unusual because two moments of the distribution, μ_n and σ_n, appear explicitly in the functional definition. One can verify that these two parameters are consistent with the earlier definitions. That is, if n is a Gaussian rv then, performing the integration required to find the expected value, one finds that

$$E[n] = \mu_n = \int_{-\infty}^{+\infty} \frac{u}{\sqrt{2\pi\sigma_n^2}} \exp\left(-\frac{1}{2}[(u - \mu_n)/\sigma_n]^2\right) du$$

and a similar result holds for the variance:

$$E[(n - \mu_n)^2] = \sigma_n^2$$

The Gaussian pdf is symmetric about μ_n, and the mean, mode, and median are all the same. There is also a simple relationship between the standard deviation and related probabilities. About 68 percent of the pdf area lies within one standard deviation of the mean μ_n, and about 95 percent lies within two standard deviations of the mean. Equivalently, a Gaussian rv has about a two-thirds chance of falling within one standard deviation of its mean, and a 95 percent chance of being within two standard deviations of the mean. Standard tables describe these probability intervals for the case of zero mean and a standard deviation of 1, which is called a standard Gaussian rv and denoted by N(0, 1). The first number refers to the mean, and the second to the standard deviation, with uppercase N meaning "normal".

The main reason for the ubiquitous use of the normal distribution is a theorem, due to Gauss, that identifies it as the likely pdf in a wide variety of situations. This central limit theorem states that the average of many identically distributed independent random variables tends to be Gaussian. This is true even if the random variables are not themselves Gaussian. They just need to be all of the same type (all realizations of an rv associated with a single pdf), and independent of one another. Independence has the colloquial interpretation that one rv knows nothing about the next or any previous rv. There is a formal definition given below: independence means that the joint pdf describing all the random variables can be factored. To mathematicians, the term "many" is equivalent to saying "in the limit of large numbers of data". To geophysicists it implies, in practical terms, that the number of data exceeds about 12.

As a result of the central limit theorem, any quantity that is likely to be the sum of many things tends to have a Gaussian distribution. When nothing is known about an rv, the Gaussian assumption is usually made. Because only two parameters are needed to specify the Gaussian pdf completely, the mean and variance, one then uses estimates of the sample mean and variance from observations; they provide a concise summary of the properties of the rv.

The central limit theorem is easy to demonstrate computationally. For example, the sum of 12 uniformly distributed rv's (described below) can be computed for say 1000 cases, and a histogram of the sum will show a bell shape. Another verification of the central limit theorem (which can be used to formulate a proof) considers the pdf of the sum of two random variables. This is the convolution of the pdf with itself. The addition of multiple random variables results in a pdf that is the multiple convolution of the original pdf. This multiple convolution tends to be bell-shaped, as we will now indicate for the discrete case of a coin toss.

In a coin toss, the random variable in each toss is independent (for a fair toss). So, the probability of obtaining zero heads in two tosses is 1/4, with the same probability for two heads. The more likely result is one head and one tail, because there are two ways to obtain it, either by first tossing a head, then a tail, or first tossing a tail, then a head. We come to the same conclusion by forming the convolution of the discrete probability series:

$$[0.5, 0.5] * [0.5, 0.5] = [0.25, 0.5, 0.25]$$

This is an ordered series, where probabilities correspond to [0, 1, 2] heads. The probability associated with the sum of four coin tosses is the convolution of the two-toss probability series with itself:

$$[0.25, 0.5, 0.25] * [0.25, 0.5, 0.25] = [0.0625, 0.2500, 0.3750, 0.2500, 0.0625]$$

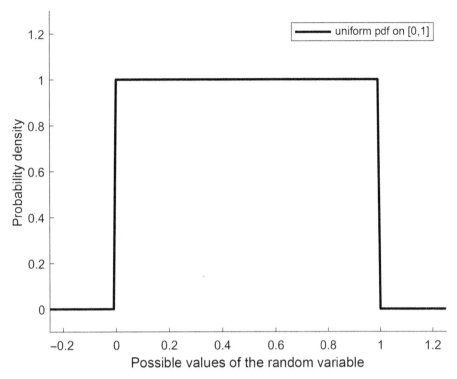

Figure C.2 The uniform pdf on [0, 1] is zero outside this interval. Most computer systems have random number generators for this random variable.

where the possible outcomes are $[0, 1, 2, 3, 4]$ heads. The probabilities associated with the outcome of eight tosses is the convolution of this result with itself. If you plotted the numbers on the right-hand side of the above equation, they would begin to form a bell-shaped curve, peaked in the middle. The actual numbers in this case are those of the binomial distribution, but the important feature is that the shape of the distribution is that of a bell, which in the limit is the Gaussian function. This discrete-probability example verifies the basic prediction of the central limit theorem. A proof of this theorem for the case of a continuous pdf can be constructed by showing that, for rather arbitrary forms of pdf, repeated convolution of the pdf with itself will yield the Gaussian function.

C.2.2 The Uniform PDF

While the Gaussian pdf arises in many physical situations, the uniform pdf (see Figure C.2) is encountered mainly in computation. Virtually all computers and operating systems have built-in algorithms to generate realizations of a random variable that is uniformly distributed. The pdf of uniformly distributed random numbers available on most computers is unity over the interval $[0, 1]$ and zero elsewhere. Uniformly distributed numbers are equally likely to fall anywhere within the interval $[0, 1]$. For this case, the associated parameters are $\mu = 0.5$ and $\sigma^2 = 1/12$. One would hope that uniformly distributed random numbers created by computer algorithms are independent of one another, so that each time the random number generator is used, it will provide a fresh realization of a uniformly distributed random variable, unrelated to values generated at other times. In some cases this is not true, because the algorithm is identical, and precisely the same set of numbers may result

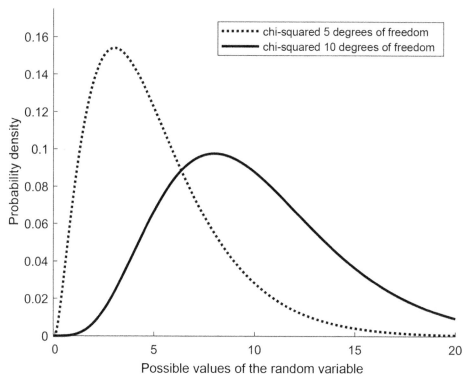

Figure C.3 Pdf for chi-squared variables with 5 and 10 degrees of freedom, created using the MATLAB pdf function. Chi-squared rv's with K degrees of freedom are the sum of K squared unit-variance Gaussian variables, so the peak moves farther to the right as the number of degrees of freedom increases.

if it is used again. The cure for this is to introduce a new "seed" to the random number generator that is random, for example, a numerical value for the time of day.

Uniformly distributed numbers are easy to generate, but are not common in physical situations. However, they can be used as the starting point for a Gaussian random number generator using the central limit theorem. If we take a sum of "many" uniform numbers, the result tends to be Gaussian, and we can scale and shift this sum to provide pseudo-Gaussian random numbers that have any desired mean and variance. For example, if we add together 12 uniformly distributed numbers (on the interval [0, 1]), and subtract 6 from the sum, the result is a zero-mean random variable that is approximately Gaussian. Each of the 12 values is independent, and as shown below this will make the variance of the sum equal to the sum of the variances. Therefore, adding together 12 uniformly distributed random numbers will be a new random variable with a variance of 1. This is an easy way to generate realizations of a standard N[0, 1] random variable.

C.2.3 The Chi-Squared PDF

The chi-squared pdf with K degrees of freedom describes the distribution of the sum of K squared Gaussian N[0, 1] random variables. It is useful in understanding the behavior and uncertainty of any quantity that is reasonably approximated as a sum of squares of random variables. Tables of chi-

Probability of being less than tabulated numerical value

dof	0.100	0.050	0.025	0.010	0.001	0.900	0.950	0.975	0.990	0.999
1	0.016	0.004	0.001	0.000	0.000	2.706	3.841	5.024	6.635	10.828
2	0.211	0.103	0.051	0.020	0.002	4.605	5.991	7.378	9.210	13.816
3	0.584	0.352	0.216	0.115	0.024	6.251	7.815	9.348	11.345	16.266
4	1.064	0.711	0.484	0.297	0.091	7.779	9.488	11.143	13.277	18.467
5	1.610	1.145	0.831	0.554	0.210	9.236	11.070	12.833	15.086	20.515
6	2.204	1.635	1.237	0.872	0.381	10.645	12.592	14.449	16.812	22.458
7	2.833	2.167	1.690	1.239	0.598	12.017	14.067	16.013	18.475	24.322
8	3.490	2.733	2.180	1.646	0.857	13.362	15.507	17.535	20.090	26.125
9	4.168	3.325	2.700	2.088	1.152	14.684	16.919	19.023	21.666	27.877
10	4.865	3.940	3.247	2.558	1.479	15.987	18.307	20.483	23.209	29.588
11	5.578	4.575	3.816	3.053	1.834	17.275	19.675	21.920	24.725	31.264
12	6.304	5.226	4.404	3.571	2.214	18.549	21.026	23.337	26.217	32.910
13	7.042	5.892	5.009	4.107	2.617	19.812	22.362	24.736	27.688	34.528
14	7.790	6.571	5.629	4.660	3.041	21.064	23.685	26.119	29.141	36.123
15	8.547	7.261	6.262	5.229	3.483	22.307	24.996	27.488	30.578	37.697

Figure C.4 Critical value tables are the standard way to make use of the chi-squared distribution. As an example, consider the case of 10 degrees of freedom, the tenth line in the table. A chi-squared random variable has a 2.5% chance of being less than 3.247, or a 97.5% chance of being greater than 3.247. On the right-hand side of the table, the same 10-degrees-of- freedom variable has a 97.5% chance of being less than 20.483. Therefore, a 10-degrees-of-freedom variable has a 95% chance of falling within the interval 3.247 to 20.483. This particular example provides bounds for a 95% confidence interval for a periodogram power spectrum estimate, as developed in Section 10.2. Numerical values are from the US National Institute of Standards and Technology (NIST).

squared random variables are therefore used in assessing uncertainty in estimates of variance and of power spectra, and in judging how well least-squares-fitted models agree with observed data. As a result of the central limit theorem, the chi-squared pdf begins to look like a normal pdf when K is larger than about 15. Chi-squared random variables are never zero or negative, and as K grows will have increasingly larger values, as shown in Figure C.3.

C.3 Multiple Random Variables

In a time series containing N samples, each may be contaminated with what is often assumed to be white noise. This is a special case in which the time series of the noise is a set of independent random variables. We need to also consider the case where the random variables may not be independent.

Independence has a simple formal definition. When two random variables are independent, the probability of the two considered together (jointly) is the product of the probability of each considered separately. For example, when a coin is tossed twice, the probability of tossing two heads in the two tosses is the product of probabilities, the probability of one head (1/2) times the probability of a second (1/2), equal to 1/4. The same definition applies to continuous random variables described by a probability density function. When two random variables are independent, the pdf can be factored into the product of two separate functions. In the common (but special) Gaussian case, two random variables are independent when they are uncorrelated (their correlation coefficient is zero).

Because it is so widely used, we describe in detail the pdf for N jointly distributed Gaussian random variables. Understanding this case is essential (and usually sufficient) knowledge for geophysicists. A jointly distributed Gaussian pdf depends only on the variances of each random variable and the correlation coefficients between every pair of random variables. Assume again that the mean of all N variables is zero, to simplify the discussion.

The joint Gaussian pdf for N random variables in a row vector $x = [x_1, x_2, \dots, x_N]$ can be expressed in terms of the $N \times N$ covariance matrix C whose elements at row i and column j are $C_{ij} = E[x_i x_j]$, the expected value of the product of the random variables x_i and x_j. The matrix C_{ij} is symmetric, and on the diagonal one finds the individual variances of each rv. The off-diagonal terms are proportional to the correlation coefficient between individual random variables,

$$\rho_{x_i x_j} = \rho_{ij} = \frac{E[x_i x_j]}{\sigma_i \sigma_j}$$

where σ_i is the standard deviation of x_i. Further discussion of correlation coefficients appears in Section C.6.3 and in Section 10.4. Covariance matrices have important uses in least squares computations, where errors in data, uncertainty in model parameter estimates, and anticipated variability in model parameter estimates are all commonly described using a joint Gaussian random variable framework.

The joint (N-variate Gaussian) pdf is

$$p_{(x_1,\dots,x_N)} = \frac{\exp(-x C^{-1} x^T / 2)}{(2\pi)^{N/2} \det(C)^{1/2}}$$

where the superscript T denotes the transpose, and $\det(C)$ is the determinant of the covariance matrix. For uncorrelated random variables (a common assumption), the covariance matrix is diagonal and the joint pdf factors into a product of simple Gaussian functions for each of the random variables in the vector x.

When the variables are correlated the pdf cannot be factored. However, the original random variables in x can be transformed to a new set of independent random variables. The inverse square root covariance matrix $C^{-1/2}$ is the transformation matrix and gives the new vector of random variables $y = C^{-1/2} x^T$. The random variables in y are all independent, because the pdf can be factored. The matrix $C^{-1/2}$ is the transformation used to weight least squares observation equations, and is similarly used in other applications where correlated random variables need to be mapped into an independent set.

It is useful to examine the bivariate case (with just two rv's, x_1 and x_2) to illustrate the basic concepts and terminology concerning jointly distributed random variables. These concepts include the definition of the marginal pdf for each rv, and the law of compound probability, which is generalized to Bayes' theorem and is the foundation of the maximum likelihood methods described in Chapter 7 and briefly here.

For simplicity, assume both x_1 and x_2 have the same variance, with zero mean. Then only two parameters are needed to specify the pdf: the standard deviation, σ_x and the correlation coefficient ρ_{12}. The expression for the N-variate case simplifies to the bivariate Gaussian joint pdf:

$$p_{x_1,x_2}(u_1, u_2) = \frac{1}{2\pi\sigma_x^2(1 - \rho_{12}^2)^{1/2}} \exp\left(\frac{-[u_1^2 - 2\rho_{12}(u_1 u_2) + u_2^2]}{2\sigma_x^2(1 - \rho_{12}^2)}\right)$$

This identifies the variables u_1 and u_2 as those that can take on any possible value of the random variables x_1 and x_2. One may verify by direct integration that

$$E[x_1 x_2] = \int_{-\infty}^{+\infty} \int_{-\infty}^{+\infty} u_1 u_2 p_{x_1,x_2}(u_1, u_2) du_1 du_2$$

is equal to $\sigma_x^2 \rho_{12}$.

The joint pdf gives probability information about both x_1 and x_2. We will now introduce an abbreviated notation in place of the pdf notation, setting $p_{x_1,x_2}(u_1, u_2) = P(x_1, x_2)$. The comma appearing in $P(x_1, x_2)$ means "and". Thus $P(x_1, x_2)$ gives the probability of both x_1 and x_2 occurring.

In addition to the joint pdf, two related statements can be made about the two random variables. One is the pdf of one variable (say x_2) without regard to the other. This is called the marginal pdf, in shorthand, $P(x_2)$. It is found by integrating over all possible values of the other variable:

$$P(x_2) = p_{x_2}(u_2) = \int_{-\infty}^{+\infty} p_{x_1,x_2}(u_1, u_2) du_1$$

The other statement is the conditional pdf, in shorthand $P(x_1|x_2)$, where the vertical bar | means "given". This is easily found from the joint pdf by first fixing the value of x_2 in the joint pdf, making it a function of u_1 alone. However, as a function of u_1, it is not quite a pdf, because it does not have unit area. To correct this situation, one divides by the integral of the joint pdf over all u_1 at the fixed value of x_2. This normalizing quantity is simply the marginal pdf, $P(x_2)$, defined in the preceding paragraph. The conditional pdf is therefore

$$P(x_1|x_2) = \frac{P(x_1, x_2)}{P(x_2)}$$

The preceding result is usually called the law of compound (or conditional) probabilities, and, since the roles of x_1 and x_2 can be interchanged, one arrives at the statement

$$P(x_1, x_2) = P(x_1|x_2) \times P(x_2) = P(x_2|x_1) \times P(x_1)$$

The right-hand two expressions can be rearranged as follows:

$$P(x_1|x_2) = \frac{P(x_2|x_1) \times P(x_1)}{P(x_2)}$$

When the rv's x_1 and x_2 are independent, $\rho_{12} = 0$ and $p_{x_1,x_2}(u_1, u_2)$ has circular symmetry in the (u_1, u_2) plane, because it depends only on $u_1^2 + u_2^2$. In that case, the marginal and conditional distributions are simple Gaussian functions, as in the univariate case. More interesting is the situation in which the two are correlated (see Figure C.5). As an example, let the true correlation coefficient be $\rho_{12} = 1/\sqrt{2}$ with equal unit variances $\sigma_x^2 = 1$. The numerical value of the correlation coefficient (discussed in Section C.6.3) implies that x_1 is N[0, 1] and that $x_2 = [x_1 + n]/\sqrt{2}$ where n is a rv that is independent of x_1 but also N[0, 1]. Both x_1 and x_2 have unit variance. The two-dimensional surface defining the joint pdf is no longer symmetric in the (u_1, u_2) plane. The marginal distribution of either variable, found by integrating along one axis or the other, will be a simple Gaussian, with zero mean and unit variance. The conditional distribution is proportional to the shape given by a cut through the joint pdf. For example $P(x_1|x_2)$ has a bell shape along a horizontal line (where x_2 is constant) intersecting the joint pdf. It is not symmetric about zero mean except for the case $x_2 = 0$.

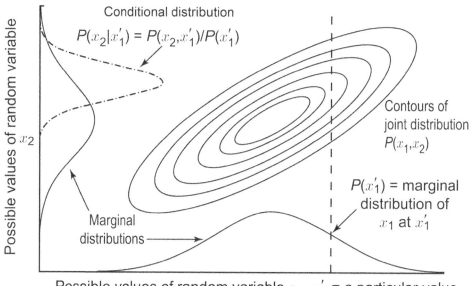

Possible values of random variable x_1 $x_1' =$ a particular value

Figure C.5   Bivariate normal pdf for variables x_1 and x_2 that are correlated. The joint distribution is a function of the two random variables, creating a surface that would stand above the plane of the page, having the appearance of an elongated hill stretching from lower left to upper right. These closed elliptical contours are schematic elevation contours like those on a topographic map, with the peak value of the joint distribution in the center of the smallest closed contour. Marginal distributions are obtained for each variable by integrating over all possible values of the other. The two marginal distributions are simple Gaussian pdfs and are plotted along each axis. The conditional distribution for x_1 is found by fixing it at some designated value, say x_1'. The conditional distribution for the other variable, $P(x_2|x_1)$, has the shape of a cross-sectional cut through the hill along a vertical plane through the dashed line at x_1 prime. The shape along the cross-sectional cut is normalized to unit area by dividing by the marginal distribution $P(x_1')$ at that value. The resulting conditional distribution is the dash-dotted curve plotted along the x_2 axis.

C.4 Bayes' Theorem and Maximum Likelihood

The law of compound probabilities was derived in the previous section from the joint Gaussian distribution but is considered to be quite general, applying to any two random variables, identified with the symbols x_1 and x_2, or sets of random variables. This liberal interpretation of the law of compound probabilities is known as Bayes' theorem. In particular, geophysical inverse methods and other theories of estimation usually associate x_1 with a model (defined by the parameters) to be estimated, and x_2 with data taken in an experiment and used to obtain an estimate. Additional discussion and applications appear in Chapter 7. Bayes' Theorem in this case is

$$P(model|data) = \frac{P(data|model)P(model)}{P(data)}$$

The quantities on the right can be computed or estimated, so the final result is a pdf for the model, which is then be then used to estimate model parameters and quantify uncertainty. In the simplest case, assume both $P(data)$ and $P(model)$ are constant, so estimates that maximize the likelihood $P(data|model)$ are chosen. These correspond to the least squares estimates developed in Chapter 7.

We can show that Bayes' theorem leads to sensible results when estimating the mean and variance of a time series n_t of length N assumed to be Gaussian stationary white noise with true mean μ_n and true variance σ_n^2. We will find an estimator, that is, a formula to give an estimate using the available data, and discuss the associated uncertainty (confidence intervals) shortly.

If we were to guess the estimator of μ_n we would probably choose, from intuition, a simple average of all data. We will conclude this is correct, but we might be curious to know whether there is a better estimator. For example, if we sorted n_t by increasing values, and took the mean of a few of the middle values, might this be better? While this question is not easily answered, we can show that the usual average maximizes the likelihood. We can use Monte Carlo experiments (generating synthetic data with a known true mean, using a random number generator) to compare this result with other proposed estimators.

Maximum likelihood proposes that $\hat{\mu}_n$ maximizes the probability of the data, given the model, $P([n_t, t = 1, 2, \ldots, N] | \mu_n)$, which, with the Gaussian assumption, is proportional to

$$\exp\left(-\sum_{t=1}^{N}[(\mu_n - n_t)/\sigma_n]^2\right)$$

This is maximized if we minimize

$$\sum_{t=1}^{N}[\mu_n - n_t]^2$$

Taking the derivative of the sum with respect to the unknown parameter μ_n and then setting it to zero leads to the estimator

$$\hat{\mu}_n = \frac{1}{N}\sum_{t=1}^{N}n_t$$

the simple average predicted by intuition.

A maximum likelihood estimator of the variance is found in a similar way to be

$$\hat{\sigma}_n^2 = \frac{1}{N}\sum_{t=1}^{N}[n_t - \hat{\mu}]^2$$

However, this estimator turns out to be biased, because the expected value of the estimate is

$$E[\hat{\sigma}_n^2] = \frac{N-1}{N}\sigma_n^2$$

The estimate is a little too small, because the mean value is not known, having been estimated from the data. In this case maximum likelihood (Bayes' theorem) leads to a reasonable but biased estimator. The fix is to multiply it by the factor $N/(N-1)$ to obtain

$$\hat{\sigma}_{n,unbiased}^2 = \frac{1}{N-1}\sum_{t=1}^{N}[n_t - \hat{\mu}]^2$$

This is the more common form of the variance estimate. For large N, as is the case in most geophysical problems, there is no practical difference. If the mean value is known rather than being estimated from the data, then the factor $1/N$ is correct.

C.5 Confidence Intervals

A standard application of normal and chi-squared random variables is to estimate intervals of confidence for estimates obtained from a finite set of data. Confidence intervals are usually stated as a percentage confidence, typically 90 or 95 percent, that the true value of an estimated quantity lies within a given interval. They are obtained using the chi-squared distribution for the variance and the Normal (Gaussian) distribution for the mean. There are many subtle arguments that underlie a full discussion. Even the idea of a confidence interval is subject to interpretation, but it is (or ought to be) something that you would bet your own money on with the stated odds. For example, if you provide a 90 percent confidence interval for some quantity, then you should be ready to bet with 9 to 1 odds that the true value of the quantity is contained in that interval. If you are not ready to do so, then the wise thing is to precede the confidence statement with the word approximate, which allows a bolder statement without full commitment. Other adjectives may also be used to indicate uncertainty in stated confidence intervals. For example, you can preface them with a qualifier like nominal or note that they are obtained under standard assumptions without having to say exactly what those are.

We review what is considered a standard way to obtain confidence intervals for mean and variance estimates for a time series n_t of length N, consisting of Gaussian independent random numbers with mean μ_n and variance σ_n^2. We assume Gaussian white noise, but the average will tend to be Gaussian for sufficiently large numbers of data for any type of random variable, by the central limit theorem. If the number of data exceeds about 12, then it is probably sufficiently large. Confidence intervals that make use of the normal (Gaussian) pdf should be valid and useful in such cases.

For the mean and variance, estimates are

$$\hat{\mu}_n = \frac{1}{N} \sum_{t=1}^{N} n_t$$

and

$$\hat{\sigma}_n^2 = \frac{1}{N-1} \sum_{t=1}^{N} (n_t - \hat{\mu}_n)^2$$

To obtain a confidence interval for the mean (assuming for the moment that the variance is known) we use the Gaussian pdf. Confidence intervals when the variance is unknown make use of a special distribution called Student-t. More about this below. The estimate of the mean

$$\hat{\mu}_n = \frac{1}{N} \sum_{t=1}^{N} n_t$$

is a new rv, the sum of N identically distributed rv's. When each n_t is Gaussian then $\hat{\mu}_n$ is also Gaussian, with the same true mean μ_n and with variance $(1/N)\sigma_n^2$. That is, the average of N Gaussian rv's is a new rv with the same mean but a smaller variance. As N gets large, the variance of the estimate diminishes, and it approaches the true mean μ_n. Further discussion of this result appears later in this appendix, in Section C.6, which considers applications of expectations and SNR improvement by averaging. Even if the members of the times series n_t are not Gaussian, but are independent, the central limit theorem predicts that their average will tend to be Gaussian.

If $\hat{\mu}_n$ is Gaussian-distributed with mean μ_n and variance $(1/N)\sigma_n^2$, then the normalized rv

$$p = \frac{\mu_n - \hat{\mu}_n}{\sigma_n/\sqrt{N}}$$

is $N(0, 1)$. Standard Gaussian (normal) tables provide numerical values for different levels of confidence. For a standard 95 percent confidence the interval is $[-2 < p < +2]$ because a Gaussian rv has 95 percent probability of falling within two standard deviations of the mean. Thus, the statement about the probability of p before the data are taken is

$$-2 < \frac{\mu_n - \hat{\mu}_n}{\sigma_n/\sqrt{N}} < 2$$

with 95 percent probability. Here μ_n is a number (unknown), and $\hat{\mu}_n$ is a random variable. After the data n_t are taken, the roles are reversed. Then $\hat{\mu}_n$ is a number computed from the data, and μ_n is a random variable. With this understanding, after the data are taken we have

$$-2 < \frac{\mu_n - \hat{\mu}_n}{\sigma_n/\sqrt{N}} < +2$$

with approximately 95 percent confidence. Then it is easy to rearrange the inequality to obtain a confidence interval for the true mean in terms of the estimate, and the variance (assumed known in this case). The result is

$$\hat{\mu}_n - 2\sigma_n/\sqrt{N} < \mu_n < \hat{\mu}_n + 2\sigma_n/\sqrt{N}$$

The confidence interval surrounds the estimate $\hat{\mu}_n$ symmetrically, and its width diminishes as N gets larger, making the estimate consistent. We can now relax our requirement that the variance σ_n^2 is known. We simply use the usual variance estimate in place of the true variance. However, now there is less information so our confidence interval may be a little too narrow, but we can inflate the interval (make it a little wider), until it seems more realistic. One easy way to do this is to reduce the number of degrees of freedom (N) by one or two, to account for estimating the variance. This approach is not exact, but it is sufficient in most cases and it emphasizes the fact that confidence intervals are really approximate. An exact confidence interval for the true mean can be established when the variance is not known, via the Student-t distribution. This distribution is distinct from the Gaussian and is often tabulated in elementary statistics books. However, it is almost never used, unless N is very small (less than 10 or so), a situation rarely encountered in geophysical problems.

The chi-squared distribution can be used to establish confidence intervals for the variance σ_n^2. The estimate $\hat{\sigma}_n^2$ is computed using $1/(N-1)$ normalization, if the true mean value is not known. Suppose $\hat{\sigma}_n^2$ has been computed from a time series of length N, using an estimated mean $\hat{\mu}_n$ in the computation. Let variable q be defined as

$$q = (N-1)\frac{\hat{\sigma}_n^2}{\sigma_n^2}$$

The variable q tends to be chi-squared distributed with $N-1$ degrees of freedom (dofs), one less than the number of data N; one dof is lost when using the estimated mean value to compute $\hat{\sigma}_n^2$. The development of a confidence interval for q makes use of a chi-squared table, which gives (usually) a set of rows for increasing numbers of dof. For each row, across the columns are values of the chi-squared variable q corresponding to varying probabilities that it will take on a value larger than the column heading. Suppose that our time series originally had $n = 10$ data, so the number of dof is 9. Then, before the experimental data are available, an interval for q is (from Figure C.4)

$$3.325 < q < 16.919$$

with 90 percent probability. After the data are taken, the variance estimate is just a number, while the true variance is an rv, so the statement is now

$$3.325 < q < 16.919$$

with 90 percent confidence. Then it is easy enough to rearrange the inequality, noting that in taking the reciprocal the signs are reversed. The result is

$$\frac{N-1}{3.325}\hat{\sigma}_n^2 > \sigma_n^2 > \frac{N-1}{16.919}\hat{\sigma}_n^2$$

When $N = 10$, this reduces to a confidence interval that is approximately given by

$$2.7\hat{\sigma}_n^2 > \sigma_n^2 > 0.5\hat{\sigma}_n^2$$

The procedure is similar to the case of the mean value, but there are two important differences. First, the confidence interval is not symmetric, as it was for the mean. The other is that the width of the confidence interval for the variance depends upon the value of the variance estimate. Confidence interval calculations for power spectrum estimates are similar, and their width depends upon the value of the power spectrum (periodogram) estimate at a particular frequency. For this reason, power spectra are best plotted on a decibel scale, so that the associated confidence intervals can be expressed in decibels relative to the estimate at each frequency, which will then be the same for all frequencies.

C.6 Applications of Expectation Operator

As noted earlier, the expected value, or *expectation*, of any function $g(n)$ of a single random variable n is found by integration using the probability density function (pdf) as a weight in the integral:

$$E[g(n)] = \int_{-\infty}^{+\infty} g(u)p_n(u)du$$

Even if we do not know the pdf, $E[\ ]$ is a linear operator (because integration is linear). Linearity allows us to come to important conclusions without knowledge of the specific pdf. Here we consider the expected values of multiple random variables, considering just two to illustrate the main conclusions and applications.

C.6.1 Variance of the Sum Equals the Sum of the Variances

If the time series of two random variables x_1 and x_2 are independent and zero mean then the variance of their sum equals the sum of their variances. This is expressed in the more succinct phrase that when two time series are random and independent, and added together, the power of their sum equals the sum of the power of each. That is, powers add, because power is proportional to variance. This result can be verified formally by finding the expected value of the variance of the sum of two random variables:

$$E[(x_1 + x_2)^2] = \int_{-\infty}^{+\infty} \int_{-\infty}^{+\infty} (u_1 + u_2)^2 p_{x_1}(u_1) p_{x_2}(u_2) du_1 du_2$$

Upon squaring, the integrand has in it $u_1^2 + 2u_1 u_2 + u_2^2$. Separating each term, integrating, and noting that the area under a pdf is unity, the first integral is the variance of x_1, and the third is the variance of x_2. The second term, which is $2E[x_1 x_2]$, is proportional to the correlation coefficient between the two. It is zero because the integral can be factored into the product of two integrals, each of which is the expected value of the random variable, which we assumed to be zero. This shows that powers add, and we can make use of this in a variety of situations where a time series representing signal is corrupted by a noise time series that is independent of the signal. Among the applications are: understanding to what extent we can improve the signal to noise variance ratio (SNR) by averaging; how to interpret and make use of correlation coefficients; and how to define a statistic, the semblance, that measures the similarity among many time series.

C.6.2 SNR Improvement by Averaging

Suppose we average (stack) two seismograms which have the same signal s_t but different random noise in each: thus the time series are $s_t + n1_t$ and $s_t + n2_t$. Assume that signal s_t is uncorrelated with the noise. The signal variance (of s_t when noise-free) is σ_s^2 and the noise variance in each seismogram is the same, σ_n^2. If $n1_t$ and $n2_t$ are completely correlated (the noise is the same for each seismogram), then averaging the two seismograms does not improve the SNR. However, if the noise in each is independent, the averaged noise will be reduced. The sum of the two independent noise series has variance $2\sigma_n^2$, the sum of their variances. At the same time the signal has been doubled, so the signal variance is $4\sigma_s^2$. Thus the ratio of the signal variance to the noise variance in the stack has improved by a factor 2, or +3 dB. Similarly, averaging N time series improves the SNR by a factor of N, or $10 \log_{10}(N)$ decibels.

Stacking is effective (and sensible) if the noise in each seismogram or other data type is independent and zero-mean, and the signal is the same. When the noise has a non-zero mean (owing to systematic errors), stacking reduces the scatter of the random part but provides no improvement in correcting systematic error. In this case the precision of a measurement is improved by stacking but the accuracy will not be improved if systematic errors are present.

Simple stacking (averaging) is sensible if the noise level in all data is about the same size. However, if some data have a lower noise level then they deserve greater weight in the stack. Really poor data can be completely discarded, that is, they can be given zero weight. Data editing (also called muting in geophysics) is used to remove obviously bad data. This is not always possible, especially when few data are available, and, even if there are sufficient data, there may still be variable data quality. In Chapter 7 on least squares we show that a weighted stack gives a sensible way to combine data of different quality. The concept of a weighted stack is generalized to allow for the possiblity that the noise among different data is correlated in a way that can be described by a covariance matrix. The inverse square root covariance matrix can be used as the weighting matrix to remove such correlations.

For the simple case of combining data of different quality, where the noise in each is unrelated, the best weights (to minimize the error variance in the stack) are the reciprocals of the noise variances, as described in Chapter 7. In this way, low-noise data are given greater weight. For example, if the noise in seismogram $s1$ has variance σ_1^2 and the noise in seismogram $s2$ has variance σ_2^2 then the optimum average is

$$\frac{s1/\sigma_1^2 + s2/\sigma_2^2}{1/\sigma_1^2 + 1/\sigma_2^2}$$

In the simple case when the noise levels are the same, the weights are just $1/2$ for each seismogram. The formula extends in the obvious way when averaging three, four, or more data.

C.6.3 Correlation Coefficients and SNR

If one assumes that two time series are linearly related, with additive noise in one or both, then the correlation coefficient depends in a simple way on the SNR. First we show how the SNR affects the correlation coefficient, and then we use the correlation coefficient to estimate the SNR. Assuming the correlation coefficient is not biased by trends or other problems, we can interpret its numerical value using an assumed model of a linear relationship between the two time series. The analysis makes uses of the expectation operator, assuming zero-mean random variables and uncorrelated signal and noise.

First, we consider the correlation coefficient between linearly related time series y and x, where x has no noise and y is noisy, so that

$$y_t = bx_t + n_t$$

We expect that if the noise n_t is zero, then the correlation coefficient should be $|r_{xy}| = 1$, with a positive or negative sign depending on the sign of b. On the other hand, if n_t has a much larger variance than the signal bx_t, the correlation coefficient is expected to be nearly zero.

The formal definition of the true correlation coefficient ρ_{xy} involves the expected values of each quantity in the definition of the Pearson correlation coefficient (the cross correlation and standard deviations). Formally this involves integration using a probability density function, but fortunately it is not necessary to know what the pdf is to get a useful result. It is only necessary to use the linearity of integration and therefore the linearity of the expected value or expectation operator $E[\ ]$ to find a result.

The definition of the true correlation coefficient is:

$$\rho_{xy} = \frac{E[x_t y_t]}{\sigma_x \sigma_y}$$

where

$$\sqrt{E[x_t^2]} = \sigma_x$$

and similarly for σ_y^2. Use of the expectation operator $E[\ ]$ implies that x_t and y_t are considered to be random variables. The ratio of the noise to signal variance is the NSR (the reciprocal of the SNR). The definition is

$$NSR = \frac{\sigma_n^2}{b^2 \sigma_x^2}$$

After simplification, setting the expectations of uncorrelated rv's to zero, the true correlation coefficient is

$$\rho_{xy} = \frac{\text{sgn}(b)}{\sqrt{1 + NSR}}$$

The sign of b determines whether the coefficient is positive or negative, and, as expected, when the NSR is large, the correlation coefficient becomes small (x and y are uncorrelated), and when the NSR is small, the true correlation coeffient approaches unity. As anticipated, the true correlation coefficient depends only upon the NSR, the ratio of the variances of the two parts of y, that is, bx_t and the part which is unrelated, n_t. Thus the NSR (or its reciprocal, the SNR) can be determined once the correlation coefficient is estimated using the usual cross correlation of the two series normalized by their standard deviations.

Another common case is to estimate the correlation between two or more noisy time series, each containing the same signal with noise at the same level (same variance) but otherwise uncorrelated. Here, calculating the correlation coefficient provides an estimate of the noise/signal variance ratio, as in the previous case, but, since now both series contain noise, the result is slightly different. The time series model in this case is

$$x1_t = s_t + n1_t$$
$$x2_t = s_t + n2_t$$

where the variance of the signal s_t is σ_s^2 and the variance of the noise for the $x1$ and $x2$ series is σ_n^2. For this case the correlation coefficient between $x1_t$ and $x2_t$ is

$$\rho_{12} = \frac{1}{1 + NSR}$$

where the noise to signal ratio, assumed to be the same in both series, is

$$NSR = \frac{\sigma_n^2}{\sigma_s^2}$$

So, again, computing the correlation coefficient provides an estimate of the NSR or its reciprocal the SNR. When the variance of $n1_t$ is σ_{n1}^2 and that of $n2_t$ is σ_{n2}^2, making the NSR different in the two series, then the correlation coefficient is

$$\rho_{12} = \frac{1}{\sqrt{1 + NSR_1}\sqrt{1 + NSR_2}}$$

If there are three different series available, then it is possible to find the NSR for each using the three possible correlation coefficients among the three series, as will be shown in the next subsection.

C.6.4 Three-Cornered-Hat Noise Estimate

When we have three different time series containing the same signal but different noise levels, we can find the SNR (NSR) in each from the three different correlation coefficients between each pair of series. This gives three equations in the three unknown NSR values, which can be arranged to find the NSR in each. We have

$$x1_t = s_t + n1_t$$
$$x2_t = s_t + n2_t$$
$$x3_t = s_t + n3_t$$

Using the three squared correlation coefficients $\rho_{12}^2, \rho_{13}^2, \rho_{23}^2$, it is easy to show that

$$1 + NSR_1 = \frac{\rho_{23}}{\rho_{12}\rho_{13}}$$

$$1 + NSR_2 = \frac{\rho_{13}}{\rho_{12}\rho_{23}}$$

$$1 + NSR_3 = \frac{\rho_{12}}{\rho_{13}\rho_{23}}$$

This allows us to determine which of the three series has the lowest noise. However, a more direct approach, allowing estimation of the actual noise variance in each series, is to take the difference between each pair of time series. The difference series is due to the noise in the two series, and its variance is the sum of the noise variances in each series, because we have assumed that the noise in each is an independent random variable and because the variance of the negative of a noise series is the same as the variance of the noise series itself. Assuming independence of the noise in each time series we can find the variance of each, and identify which series is of the best quality, with the lowest noise level. This is called a three-cornered-hat estimate because it involves three or more sets of observations.

Suppose the three series are $yk_t = x_t + nk_t$ for $k = [1, 2, 3]$, where nk_t is the independent noise series in each with variance σ_k^2. The variance of the differences among the three series are σ_{12}^2, σ_{23}^2, and σ_{13}^2. Since the variance of the sum is the sum of the variances for independent noise, we have $\sigma_{12}^2 = \sigma_1^2 + \sigma_2^2$, and so on for the other two differences. Then the variance in each series is found by solving the following simultaneous linear equations:

$$\begin{bmatrix} 1 & 1 & 0 \\ 0 & 1 & 1 \\ 1 & 0 & 1 \end{bmatrix} \begin{bmatrix} \sigma_1^2 \\ \sigma_2^2 \\ \sigma_3^2 \end{bmatrix} = \begin{bmatrix} \sigma_{12}^2 \\ \sigma_{23}^2 \\ \sigma_{13}^2 \end{bmatrix}$$

If more than three time series are available, a solution may be obtained by expanding the equations to include additional difference series and solving the equations via least squares. The three-cornered-hat method for finding SNR can be also adapted to finding noise variances as a function of frequency. The power spectrum is a narrow-band estimate of variance, so the calculation would be performed at each frequency using the power spectra of the difference time series.

C.6.5 Semblance

When the correlation among multiple time series is to be measured, a joint statistic called the semblance is used. The semblance is the normalized ratio of the variance of the sum of all the time series and the sum of the individual time series' variances. For example, for three zero-mean time series x_t, y_t, z_t, their semblance is

$$S_{xyz} = \frac{1}{3} \frac{\sum_{t=0}^{N-1}(x_t + y_t + z_t)^2}{\sum_{t=0}^{N-1}[x_t^2 + y_t^2 + z_t^2]}$$

The formula generalizes to the case of four, five, or more series. The numerator of the semblance statistic is the variance of the sum or stack of the time series. The denominator is the sum of the individual variances. When the time series are uncorrelated, the variance of the sum equals the sum of variances, since this is the case of uncorrelated random variables. Thus unrelated time series have a semblance equal to $1/3$, in this case, and more generally to $1/K$ for K time series that are unrelated. Semblance calculations are sensible for all time series that have been normalized to have about the same variance, preconditioned to remove trends, and prefiltered to have an approximately white spectrum. Then values significantly larger than $1/K$ indicate correlation or common features among all the time series.

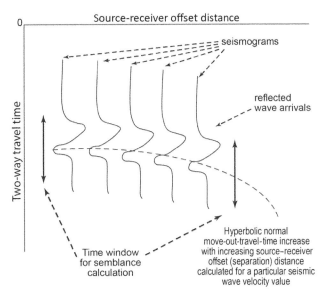

Figure C.6 A schematic set of reflection seismograms taken at increasing source–receiver offsets. A particular reflection event appears on each, but at increasing travel time owing to increasing source–receiver separation. This increase in travel time is called normal move-out, and its detailed shape can be used to estimate the velocity structure between the surface and the depth of the reflector producing the event. The process is automated by using the semblance among all seismograms, computed for a time window within each seismogram. In the automated process, known as computation of a velocity spectrum, one tests trial velocity values to find those which best predict the observed normal move-out. This approach both provides an estimate of velocity and also determines the time shifts needed to align the seismograms so that they can be stacked (added together) to improve the signal to noise ratio.

An important application of semblance (as defined here, or a similar statistic) in geophysical data processing is in stacking or averaging seismograms, while at the same time estimating seismic velocity. The data are many seismograms generated by a seismic source (an explosion or Vibroseis) recorded at different source–receiver separation distances. Figure C.6 depicts schematically how seismograms (in practice, often many hundreds) are to be stacked to improve the SNR. See the discussions of SNR improvement from stacking earlier in this appendix. The increase in travel time with source–receiver offset can be predicted for a given seismic velocity and defines a hyperbolic curve. For a particular travel time window, indicated in the figure, a range of trial velocities is tested and the semblance among the move-out-corrected seismograms (that is, they have been corrected for increasing travel time) is computed for each trial value. The largest semblance value identifies the best estimate of velocity and allows the seismograms to be aligned so that they can be stacked.

Appendix D **Further Reading**

There are many books and journal articles on the subjects in this book, and also many resources available on the internet. This list includes books that cover related material, often in a more advanced form. These are arranged in author alphabetical order.

Bloomfield, Peter: *Fourier Analysis of Time Series: An Introduction*, Wiley Series in Probability and Mathematical Statistics, 1976.

Bracewell, Ron: *The Fourier Transform and Its Applications, Second Edition*, McGraw Hill, 1978.

Chatfield, Chris: *The Analysis of Time Series: An Introduction*, Chapman and Hall/CRC, 1984.

Claerbout, John: *Fundamentals of Geophysical Data Processing, With Applications to Petroleum Prospecting*, McGraw Hill, 1976.

Gubbins, David: *Time Series Analysis and Inverse Theory*, Cambridge University Press, 2012.

Hamming, R.W.: *Digital Filters, Second Edition*, Prentice Hall, 1983.

Karl, John H.: *An Introduction to Digital Signal Processing*, Academic Press, 1989.

Menke, William: *Geophysical Data Analysis: Discrete Inverse Theory, Fourth Edition*, Academic Press, 2018.

Oppenheim, Alan and Schafer, Ronald: *Discrete-Time Signal Processing*, Prentice-Hall, 2009.

Orfanidis, Sophocles: *Introduction to Signal Processing*, Prentice Hall Signal Processing Series, 1996.

Ramirez, Robert: *The FFT, Fundamentals and Concepts*, Prentice Hall, 1985.

Robinson, Edward: *Data Analysis for Scientists and Engineers*, Princeton University Press, 2016.

Yilmaz, Oz: *Seismic Data Analysis: Processing, Inversion, and Interpretation of Seismic Data*, Society of Exploration Geophysicists, 2001.

Index

acausal filter, 50
aliasing, 12, 14, 15
all-pole filter, 96
all-zero filter, 96
ambient-noise seismology, 129
analog signal, 11
analog to digital converter (ADC), 11
analytic signal, 33, 167
angular frequency, 25
anti-alias filter, 13
arctangent computations using ATAN2, 30
Argand diagram, 30
ARMA filter transfer function, 56
associative property of convolution, 52
autoregressive filter, 50
autoregressive moving average (ARMA) filter, 50
autoregressive spectrum estimate, 124
autocorrelation, 17
autocorrelation for the Fourier transform, 159
autocorrelation theorem, 63

back-loaded couplet, 97
Backus filter, 108
band-limited interpolation, 43
band-limited signals, 12
Bartlett window, 64
Bayes' theorem, 75, 178
Bayesian inference, 76
bilinear transform, 102
blind deconvolution, 123
Bode plot, 55
boxcar function, 161
boxcar window, 64
Burg algorithm, 119

Cartesian form of a complex number, 30
causal filter, 50
central limit theorem, 73, 172
chi-squared pdf, 171
chi-squared random variable, 138
chirped radar and sonar, 124
circular convolution theorem, 62
circumflex denoting an estimate, 16
coherence spectrum, 146
column space of a matrix, 79
commutative property of convolution, 52
compact disc (CD), 21
complex conjugate, 30
complex Fourier series, 37

complex frequency, 34, 113
conditional pdf, 177
conditioning a time series, 141, 146
confidence interval, 180
confidence intervals for periodogram estimates, 139
convolution and correlation for the Fourier transform, 158
convolution and polynomial multiplication, 52
convolution matrix, 53
convolution theorem, 52
correlation and convolution, 53
correlation coefficient, 53, 145
correlation coefficient and SNR, 184
correlation filter for precise timing, 127
couplet, 97
covariance matrix of data errors, 176
covariance matrix of data errors C_d, 81
covariance matrix of estimated parameter errors C_m, 82
Crank–Nicolson finite difference scheme, 104
critical sampling, 13
cross correlation, 53
cross correlation for the Fourier transform, 159

damped least squares, 117
Darwin tidal notation, 84
data editing, 81
data trimming, 81
DC component, 25
de-aliasing, 12
de-blurring filter, 130
decibel (dB), 20
decimation, 14
deconvolution filter, 123
delta function, 162
derivative filter, 104
derivative theorem for the Fourier transform, 160
digitizing analog signals, 12
discrete convolution, 51
Discrete Fourier Transform (DFT), 38
Discrete Fourier Transform normalization, 40
Discrete Fourier Transform zero-padding, 43
discrete prolate spheroidal wave function, 67
discrete transient convolution, 52
dispersion of surface waves, 46
distributive property of convolution, 52
dot product of vectors, 152
downward continuation of gravity field, 112
dynamic range, 19

echo filter, 105
Euler relationships, 31
even function, 39, 157
even time series, 39
expectation operator, 170, 182
expected value, 170, 182
explained variance, 91

false white noise, 117
Fast Fourier Transform (FFT), 42
feedback terms, 51
filter cascade, 57
finite impulse response (FIR), 57
FIR filter, 57, 96
floating point variable, 19
forward transform, 39
Fourier frequency, 26
Fourier series to Fourier transform, 156
Fourier transform notation, 156
front-loaded couplet, 97

gapped prediction, 119
Gaussian pdf, 171
geometrical interpretation of least squares, 79
geostatistics, 120
ghost filter, 105
Gibbs oscillations, 29, 45
Global Positioning System (GPS), 127
GPS pseudo-random binary code, 127
gravity anomaly calculation, 109
group velocity, 46

half-power point, 21
Hanning window, 64
Heaviside function, 166
Hermitian symmetry, 37, 96
hertz unit (Hz), 11
Hilbert transform, 44, 166

identity matrix, 152
imaginary unit, 29
impulse response l_t, 57
impulse time series δ_t, 57
infinite impulse response (IIR) filter, 57, 96
integer variable, 19
integration filter, 104
interpolation, 14
interpolation filter, 120
interpolation with the DFT, 43
inverse Discrete Fourier Transform (IDFT), 38
inverse filter, 58
inverse theory, 74
inverse transform, 38

joint Gaussian pdf, 176

kriging, 120

lag, 17
lag-zero autocorrelation, 17

least squares, 74
least squares inverse filter, 116
line spectrum, 141
linear filtering with the DFT, 68
Love waves, 46

mantissa, 19
marginal pdf, 177
matched filter, 124
matched-Z transform, 102
matrix compatibility for multiplication, 152
matrix inverse, 152
matrix multiplication, 152
matrix representation of simultaneous linear equations, 152
maximum delay couplet, 97
maximum entropy spectrum estimate, 124
maximum likelihood, 74, 178
maximum-phase couplet, 97
mean value, 15
median, 16
metastable filter, 57
Milankovitch periods, 147
minimum delay couplet, 97
minimum-phase couplet, 97
mixed-pole-zero filter, 96
modulus of a complex number, 30
moment of a random variable, 170
moving average (MA) filter, 50
moving average filter coefficients from sampled impulse response, 108
multi-taper method, 67
multiple random variables, 175

negative frequency, 32
noise to signal ratio (NSR), 184
normal equations, 78
normal move-out, 186
normal pdf, 171
notch filter, 98
Nyquist band, 32
Nyquist frequency, 13
Nyquist–Shannon sampling theorem, 12

observation equations, 77
ocean tide prediction, 84
odd function, 39, 157
odd time series, 39
order of a digital filter, 51
orthogonality, 28
outliers, 81
over-determined equations, 77, 152
oversampling, 13

parametric spectrum estimate, 124
Parseval theorem, 43, 137
partial Fourier sums, 29
period, 25
periodogram of white noise, 137
periodogram spectrum estimate, 136
phase, 25

polar reciprocal, 97
positive frequency, 32
power, 16
power spectral density (PSD), 12, 137
power spectrum of a continuous function, 165
power transfer function from Z polynomials, 97
prediction error filter (PEF), 122
prediction error filter spectrum estimate, 124
principal alias, 15
probability density function (pdf), 169
probability of the sum of random variables, 173
prolate spheroidal wave function (PSWF)
    window, 67
pure-imaginary number, 30
pure-real number, 30

Rayleigh theorem, 166
record length, 26
regression, 74, 77
reverberation filter, 105
rms, 16
root mean square (rms) value, 16

sample mean, 16
sample variance, 16
sampling frequency, 12
sampling theorem, 12, 13
sampling theorem proof, 163
satellite radar altimetry, 88
sea level, 4
seismic tomography, 86
semblance, 186
shift theorem for the Fourier transform, 160
sign function, 166
signal attribute, 33
signal conditioning, 13
signal to noise ratio (SNR), 183
signal to noise ratio improvement by averaging, 183
significand, 19
similarity theorem for the Fourier transform, 160
sinc function, 161
single precision, 19
sinusoidal coefficients from the DFT, 40
sinusoidal fitting with least squares, 83
Slepian functions, 67
slowness, 86
spatial frequency, 25
spectral factorization, 97
stable filter, 57
standard deviation, 16

stationary, 15
statistic, 15
Student-t distribution, 181
sum of squares of errors, 76

temporal frequency, 25
three-cornered-hat noise variance estimate, 185
tidal admittance, 85
tidal constituents, 84
tide prediction machine, 85
Toeplitz matrix, 117
transfer function $L(f)$, 54
transfer function computation using DFT, 56
transfer function ratio of Z polynomials, 56
transpose of a matrix, 152
trapezoidal rule integration filter, 104

under-determined problem, 80
undersampling, 14
uniform pdf, 171
uniform sampling, 12
unit circle, 31
unit modulus complex number, 31
unstable filter, 57
unwrapped phase, 45
upward continuation of gravity field, 112

Vandermonde matrix, 43
variance, 16
variance estimate bias, 179
velocity spectrum, 186
Vibroseis, 124
voxel, 86

wavenumber, 25
weighted average, 82
weighted least squares, 81
white noise, 18
whitening filter, 123
Whittaker's interpolation formula, 164
Wiener filter, 116
window function, 64

Yule–Walker equations, 117

Z plane pole, 96
Z plane zero, 96
Z transform convolution theorem, 52, 61
Z transform defined, 42
zero-phase filter implementation, 100